To Catch the Rain

Inspiring stories of communities
coming together to catch their own
rain, and how you can do it too.

Lonny Grafman

Humboldt State University Press

HSU Library
1 Harpst Street
Arcata, California 95521-8299

hsupress@humboldt.edu
www.tocatchtherain.org

Cover design by Lonny Grafman
Cover image by Brians_Photos licensed Creative Commons CC0
Interior design and layout by Marian Voicu of Layouts.ro
Diagrams by Gabriel Krause
Editing by Lori Snyder of Yoga:edit
Consulting provided by Mikaylah Rivas
Reviewing and advising provided by many Appropedians

ISBN 13: 978-1-947112-04-9
Library of Congress Control Number: 2018930066
10 9 8 7 6 5 4 3 2 First Print

Lonny Grafman
with custom illustrations by Gabriel Krause
and images and inspirations
from dozens of Appropedia contributors

Dedicated to all the
amazing, wild, and engaged
future-makers
whom I have the luck to ride with.

Table of Contents

Table of Figures

Table of Equations

Foreword

To Catch the Rain by Lonny Grafman leads with the motivating sentence "Inspiring stories of communities coming together to catch their rain and how you can do it too." This guidebook is meant for practitioners and partners, whether students, teachers, designers, engineers, organizers, or end users.

Bringing his experiences from past collaborations and the plans of dozens of Appropedia users together, *To Catch the Rain* shares a selection of different strategies for building low-cost rainwater catchment systems with technologies that are appropriate for the place, the available materials, and the amount of rainwater available for capture. In practice, the nuts and bolts of rainwater catchment in Section 2 is an invaluable resource as it provides components, tools, working methods, and a rare window into the cooperative planning and building processes.

To Catch the Rain highlights the importance of community-based participation throughout each part of a design and development process, beginning with the important step of inclusively identifying the project to be undertaken. Grafman's comprehensive writing allows people to grasp the subject from different experience levels. At the core of his argument is that working together allows us to find the best solutions as well as multiple workarounds to overcome obstacles. *To Catch the Rain* is our satellite roadmap. With it, we can identify three or four different approaches to reach our destination. One approach may contain traffic, one may contain tolls, and one may be more scenic. Each way can get us there, and our criteria will inform the route we choose.

– **Mary Mattingly** co-creates sculptural ecosystems in urban spaces. Currently, Swale is a floating food forest in New York City, and she recently completed "Pull" for the International Havana Biennial. Mattingly founded the Waterpod Project, a barge-based public space and self-sufficient habitat. In 2009 the Waterpod hosted over 200,000 visitors in New York. Her work has been featured in Art in America, Artforum, Grey Room, Sculpture Magazine, China Business News, The New York Times, Financial Times, Le Monde, New Yorker, The Wall Street Journal, on BBC News, MSNBC, Fox, NPR, WNBC, and Art21.

To Catch the Rain is targeted at makers – DIYers looking for practical solutions to water problems for themselves or their communities. The book also provides an excellent overview of rainwater harvesting for students and teachers in environmental science, sustainable design, international development and engineering.

Lonny Grafman, the founder of the world-renowned user-developed website Appropedia, takes the reader through a tour de force of real, practical, global experiences of rainwater catchment at both the individual and community levels. Grafman walks the reader step-by-step through transparent design analysis to successfully engineer rainwater harvesting systems that work whether you are in one of the most remote impoverished places in the world or in wealthy and technically savvy California.

He shares intriguing stories from dozens of locations and synthesizes the work of hundreds of engaged community members and students that have contributed to the open source appropriate technological development of rainwater harvesting. *To Catch the Rain* is richly illustrated with pictures of real rain harvesting systems from all over the world – both what works and what doesn't. Intriguingly, the failures and solutions can be the most useful and thought provoking. The end of the book provides detailed stories on a wide selection of rain catchment systems and the people that made them. What the systems all have in common is careful attention to being socially and culturally appropriate for their users. For example, Grafman ingeniously illustrates how his team purposefully rigged one of their systems to break whenever students in the school attempted to use it to climb to the roof. A simple zip tie fixes the break while ensuring the integrity of the system and keeping students off the roof of the school! Although Grafman and his students worked on many of the systems highlighted in *To Catch the Rain*, much of the information came from the global Appropedian community. It is both touching and empowering to see people helping one another by sharing what works for providing rainwater for practical engineered use like the simple act of drinking a clean cup of water. This is a must read for anyone that is being denied that simple pleasure.

– **Dr. Joshua M. Pearce** developed the first Sustainability program in the Pennsylvania State System of Higher Education and the Applied Sustainability graduate engineering program while at Queen's University, Canada. He currently is a Professor cross-appointed in the Department of Materials Science & Engineering and in the Department of Electrical & Computer Engineering at the Michigan Technological University where he runs the Open Sustainability Technology Research Group. His research concentrates on the use of open source appropriate technology to find collaborative solutions to problems in sustainability and poverty reduction. He is the editor-in-chief of HardwareX, a journal dedicated to open source scientific hardware and the author of the *Open-Source Lab: How to Build Your Own Hardware and Reduce Research Costs.*

Figure 1
Community meeting in the Dominican Republic to assess needs and resources. Notice the flipchart paper taped to the walls and floor for each break-off group.

Preface

I have been incredibly lucky. Every day, I have the pleasure of working with engaged community members around the world to use available resources to meet pressing needs (Figure 1). This book highlights just a few of those projects, the ones that focused on catching the rain. Community-based design creates the lens through which this book is written, and it provides the basis for the examples and stories included.

In 2005, shortly after starting to teach at Humboldt State University, I co-created a university summer abroad program in Northern Mexico called Practivistas. Actually, the program remained nameless for years, until a fateful meeting with a new community partner finally gave us a name (more on that later). This program eventually moved to the Dominican Republic and became the Practivistas Dominicana program. In the Practivistas Dominicana program, students primarily from the US travel to the Dominican Republic, where they live with a family of their own and study at Universidad Iberoamericana (UNIBE) with local students. Together the local and foreign students work in financially poor communities to design, build, learn, and implement appropriate technologies.[1] The students come not as tourists, but as co-conspirators. The students come not with solutions, but with an earnest desire to discover solutions together. Everyone brings the resources and know-how they have to bear; for example, a student may come with internet access and research skills, whereas a community member may come with building skills. The students come with fresh eyes, and the community members come with a profound knowledge of the existing systems.

---

1   Appropriate technology is an ideological movement that involves small-scale, labor-intensive, energy efficient, environmentally sound, people-centered, and locally controlled projects. Hazeltine, B.; Bull, C. (1999). Appropriate Technology: Tools, Choices, and Implications. New York: Academic Press. pp. 3, 270. ISBN 0-12-335190-1. For me they are technologies and processes that meet current needs with available resources, while bolstering local capacity and improving the future. In addition, they are designed for and with the people using them.

Together we start with an open community process, prioritizing needs and discovering resources. We start with no idea what we are building, besides friendship and trust, and work our way from idea to a finished product, all in just six weeks. In this way we have built renewable energy systems and workshops, plastic bottle (ecoladrillo) schoolrooms, waste material blocks (hullkrete), waste plastic extruders, bamboo community structures, and, the subject of this book, rainwater catchment systems (Figure 2).

Figure 2
Grade school and university students working together to install and inspect a rainwater storage tank for a school in Dominican Republic... and having a good time!

In 2006, I founded Appropedia.org as a place for community members, students, and practitioners to share solutions and failures related to sustainability, appropriate technology, international development, and poverty reduction projects. Quickly, our community grew, and with a few deeply engaged colleagues such as Chris Watkins, Curt Beckmann, and Cat Laine, we co-founded The Appropedia Foundation as a 501(c)3 nonprofit.[2]

In the eleven years following, Appropedia grew to thousands of pages and developed a reputation for the nitty-gritty content that could not be found anywhere else. As of this writing, Appropedia has drawn over 75 million views from all around the world. Even more inspiring to me is that we have had over 350,000 edits from our engaged members.

I find a special joy when I am making a site visit for an appropriate technology project and find out that they used Appropedia to learn how to make it. In 2009, I was checking out a well-built rainwater catchment system in Nicaragua that had a first-flush. A first-flush is a vital, but not often known to be needed, component of a catchment system (more on this later). I asked how they knew to build a first-flush, and they replied that they found it on the internet. Hiding my excitement, I asked where on the internet... and they said Appropedia!

Those little moments highlight the importance of sharing. What keeps me excited and optimistic is remembering how much knowledge we all hold to share with each other. Instead of rebuilding the same proverbial wheel, we can build better wheels and advance real solutions. Although there is no panacea for our problems, there are countless solutions all around the world ready to be implemented, innovated, and improved.

---

2   Pearce, Joshua M., Lonny Grafman, Thomas Colledge, Ryan Legg, (2008) *Leveraging Information Technology, Social Entrepreneurship, and Global Collaboration for Just Sustainable Development.* 12th Annual NCIIA, 201-210.

Figure 3
Community meeting in Las Malvinas, Dominican Republic.

I also have the honor of teaching sustainable design, energy, and appropriate technology at Humboldt State University (HSU). Humboldt State University is located in Northern California and is home to many appropriate technology endeavors, such as the Campus

Center for Appropriate Technology, the Arcata Marsh, the Schatz Energy Research Center, and Potawot Health Village. Surrounded by six rivers and the Pacific Ocean, we give water a prevalent role in our curriculum.

My classes at HSU are service-learning classes, working with local, domestic, and international partners to tackle real problems. Together we have built hundreds of solutions. Most of these solutions are still having an impact and are documented on Appropedia with how-to details and follow-ups years after implementation. Appropedia saves us from re-inventing the wheel, and service learning saves us from just spinning our wheels. Students have real impact and all of the learning has immediate context.

Fourteen years of rainwater projects developed with students and communities fill the following pages with hands-on, practical experience of rainwater catchment components, systems, mathematics, and real stories. I hope you enjoy, learn, and share.

Free as in Stone Soup and Liberty

The projects described here were created by hundreds of engaged community members and students. I feel so lucky to have played some part in most of them. The images are all licensed as Creative Commons – Share Alike – By Attribution (CC BY-SA). That means that you are free to reuse and repurpose any of this content in any manner you like, as long as you attribute the work and allow others to share as well. In addition, proceeds from the purchase of the physical book go directly to the Appropedia Foundation.

Appropedia is the wiki for sustainability run by the non-profit 501(c)3 Appropedia Foundation, which is dedicated to:

Sharing knowledge to build rich, sustainable lives.

APPROPEDIA
the sustainability wiki.

1. Introduction

If water is life, rainwater is a fountain of life. Rainwater is renewal. Each trip around the hydrological cycle brings a fresh start, clean water, and a new chance for health and life. Catching the rain allows us to store that clean water and thrive as communities. To catch the rain is to believe in the future.

The purpose of this book is to share inspiring stories and empowering knowledge about how various communities have caught that fountain of life using rainwater harvesting systems. This book looks at real, practical, global experiences of rainwater harvesting (aka rainwater catchment) at individual and community-based levels through academic, mathematical and practical perspectives. This book can be used to learn practical skills, hear real stories, and make math have more meaning.

This book is for practitioners, DIYers, and community members looking for water solutions, as well as for students and teachers in environmental science, environmental studies, sustainable design, international development, engineering, and mathematics. There are sections on rainwater harvesting in general, types, components, gravity, calculations, stories, useful links, conversions, and problem-sets. If you are looking for inspiration, jump to the stories.

The calculations in this book are intended to take a rather general approach. Rules of thumb as well as deeper calculations are used. These calculations can be applied both to unique rainwater harvesting applications as well as systems beyond rainwater. The questions at the back of the book are designed for a wide range of classes.

When I travel and lecture on rainwater harvesting, I always start with the same question: *"Who owns the rain?"* That question has no single answer and often yields creative, sometimes stunningly accurate responses, such as nature, everyone, no one, the animals, Bechtel, the people, the city, the government, Nestle, whoever used it first, the river, and more.

The next question I ask is, "*Why catch rainwater?*" My assumption is that if you are reading this book, you already have an idea of why. That said, here are a few reasons:

- Water security and resilience
- Less impact on the wastewater treatment plant
- Water quality
- Fewer chemicals
- Cost
- More connection to resources
- Independence

- Less energy
- Less impact on often failing sewer systems
- Less runoff, less erosion
- Less processing
- Local groundwater recharge
- Because it works
- Less waste

In some communities, the reasons are glaringly apparent – people, mostly children, are dying from preventable waterborne illness. In other communities, the reasons are more personal – a deep and personal care for the environment, and/or artistic, aesthetic, or whimsical motivations. Often when people learn of the energy and chemical processes embedded in delivered water, seeing rainwater go unused invokes a greater sense of waste.[3] For some families, it is a desire to connect their family to their natural resources and consumption. In all these cases, catching rain builds local and personal capacity and agency to build a more secure future together. The next sections lead you through how to do it and share personal stories of how it has been done.

---

3 The Japanese word *Mottainai* refers to that sense of regret one feels when seeing something wasted. I love that having a word for it seems to make the feeling even more valid.

1.1 Rainwater Harvesting

The basic concept of rainwater harvesting is to catch the water falling from the sky and use it with more direct purpose and before it gets dirtier from the ground. Section 1.2 covers the main types of systems, Section 2 covers the components of a rainwater harvesting system, Section 3 covers gravity and pressure, and Section 4 dives into the calculations for sizing various aspects of a rainwater harvesting system.

Section 5 describes community engagements that built the types of systems covered in this book, while Section 5.7 covers some organizations implementing this knowledge. The background and history of rainwater harvesting is rich, and out of the scope of this book. If you are interested in learning more, please see the resources in Section 6. Finally, Section 7 contains problem sets to test your knowledge and prepare you for designing your own unique system.

1.2 Rainwater Harvesting Types

Rainwater harvesting is a highly customizable technology. Myriad solutions exist across a global context. Rainwater harvesting systems can usually be categorized as either passive or active types. A passive type uses almost no mechanical means to capture, convey, or treat the caught rainwater. An active type uses mechanical and/or electrical means to capture, convey, and/or treat rainwater.

I often avoid using the terms *passive* versus *active* and instead refer to *landscape* versus *built* types. Landscape type systems of rainwater harvesting, as shown in Figure 1-1, use landscape features to slow, absorb, and/or store rainwater. Landscape type systems are usually considered passive.

Figure 1-1

Landscape (passive) rainwater systems. A sand dam (left - appropedia.org/Sand_dams) in a periodically wet river bed and a bioswale (right - appropedia.org/Potawot_swales) off a parking lot in a wet climate provide retention and erosion prevention. Diagram by Gabriel Krause.

A built type system, as shown in Figure 1-2, uses mechanical and/or electrical means to capture, convey, and/or treat rainwater. Built type systems are usually considered active.

Landscape types have the advantage of lower cost per volume of water. Built types have the advantage of providing cleaner water. Many systems incorporate aspects of both, for example conveying the overflow from a built tank into local landscaping.

Figure 1-2

Built (active) rainwater systems. A simple wine barrel and downspout (left - appropedia.org/Sunny_Brae_rainwater_catchment_system) in Humboldt County, USA and a household system (right - appropedia.org/Rainwater_catchment_at_Isla_Urbana) in Mexico City.

Built type rainwater harvesting systems can be further categorized into dry versus wet systems (Figure 1-3). A dry system is one in which the conveyance system (e.g., downspouts) is evacuated and dry between rains, whereas a wet system is one in which the conveyance system remains filled with water between rains. A dry system often has pipe overhead, whereas a wet system often has pipe below ground. A dry system has the advantage of less clogging and fewer insects. A wet system has the advantage of being able to pipe the water underground and back up into a tank to keep the piping out of the way.

Dry versus wet rainwater harvesting system.

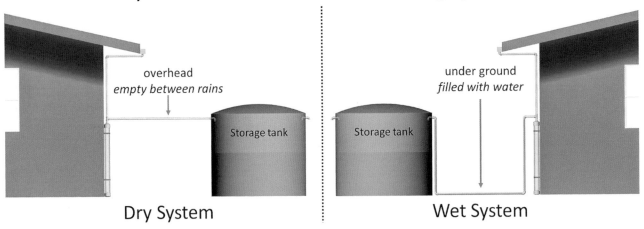

Figure 1-3
Dry versus wet rainwater harvesting systems. Diagram by Gabriel Krause.

A characteristic important to all rainwater harvesting systems is the source of pressure. Pressure is what moves water from one point to another. The source of pressure can be gravity or pumps. Gravity systems have the benefit of lower cost, lower maintenance, and being generally more robust, but they rely on elevation change. Pump systems have the benefit of being able to generate the needed pressure independent of available terrain, but are higher cost and require energy input.

A final important aspect of rainwater harvesting systems is their scope and/or size. Rainwater harvesting systems can serve a small garden, a house, a school, a community, or an even larger project. The larger the scope and size, the more costs are involved and the more water can be utilized.

This book focuses primarily on built, dry, gravity-fed systems for garden, household, and small organizations.

2. Components

Built rainwater harvesting systems use various components to best meet needs. These components can be broken down into catchment surface, conveyance (gutters and downspouts), screens, first-flush, storage, water purification, and end use (Figure 2-1).

1. **Catchment surface** – area that the rainwater falls on to be captured.

2. **Conveyance** (gutters and downspouts) – transports the water from catchment to storage or use.

3. **Screens** – separates debris from the water.

4. **First-flush** – diverts the first, and dirtiest, portion of rainwater.

5. **Storage** – holds water for later use.

6. **Purification** – cleans the water to the needed level.

7. **End use** – gives purpose to the system!

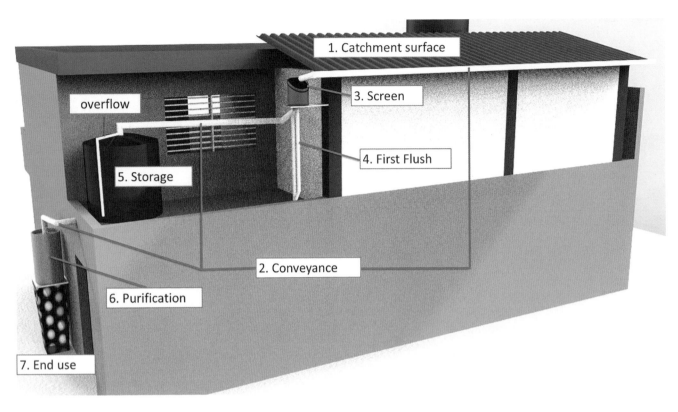

1. Catchment surface

3. Screen

overflow

4. First Flush

5. Storage

6. Purification

2. Conveyance

7. End use

Figure 2-1
The system at La Yuca, with labels for each of the main components : catchment surface, conveyance (gutters and downspouts), screens, first-flush, storage, purification, and end use. appropedia.org/La_Yuca_rainwater_catchment_2011

2.1 Catchment Surface

The catchment surface is the area that the rainwater falls on to be captured. The catchment surface in a built (active) system is typically a roof.

The main questions when designing a catchment system are usually the following:

- When can we catch the rainwater (i.e., what are the seasons)?

- How much rainwater can we catch?

- How much water do we need?

- What can we afford in money (initial, operation and maintenance, etc.)?

- What are our overall goals?

- What can we afford in time (initial, operation and maintenance, etc.)?

- What expertise and equipment are accessible?

- What do we need it for?

- What are the social, cultural, political, and other appropriate constraints?

The catchment surface is important to some of the main questions, e.g., to determine how much water you can catch, how the water can be used, and what end uses are appropriate. The surface material and size is part of determining the total catchable volume. The surface material also affects what the water can be used for, as shown in Table 2-1. The surface height is used to determine where you can use the water or whether you will need a pump.

Some typical roof materials are listed in the following table (Table 2-1):

Table 2-1
Rainwater Harvesting Surfaces

Roof material[4]	Catchment potential[5]	Water safety[6]
Thatch	Low	Low
Tile	Medium	High
Asphalt	Medium-High	Low
Concrete	Medium-High	Medium
Corrugated metal	High	Depending on coating
Standing seam metal	High	High

Roof material impacts are still being studied, and in specific cases, where purity is critical, it is suggested that water be tested periodically[7] for various pollutants and environmental data such as pH, BOD, TSS, fecal coliforms, etc. The following images show various roof materials such as asphalt shingles (Figure 2-2), concrete (Figure 2-3), clay tiles (Figure 2-4), and metals (Figure 2-5 and Figure 2-6).

4   Mendez, C. B., Afshar, B. R., Kinney, K., Barrett, M. E., & Kirisits, M. (2010, January). Effect of Roof Material on Water Quality for Rainwater Harvesting Systems. Retrieved from https://greywateraction.org/wp-content/uploads/2014/11/Effect-of-Roof-Material-on-Water-Quality-for-Rainwater-Harvesting-Systems.pdf

5   Porter, D.O., Persyn, R.A., Silvy, V.A. (2004). Rainwater Harvesting. Fort Stockton, TX. Texas A&M University.

6   DeBusk, K., & Hunt, W. (2014, February). Rainwater Harvesting: A Comprehensive Review of Literature. Retrieved from https://repository.lib.ncsu.edu/bitstream/handle/1840.4/8170/1_NC-WRRI-425.pdf

7   Tests can be done at local labs. Open source and DIY methods are listed at appropedia.org/Water_quality_testing

Figure 2-2
Asphalt shingles and downspout to wine barrel in California, USA. The water from asphalt shingles is considered too dirty for potable use. This system is appropriately designed for toilet flushing and watering landscape plants. appropedia.org/Sunny_Brae_rainwater_catchment_system

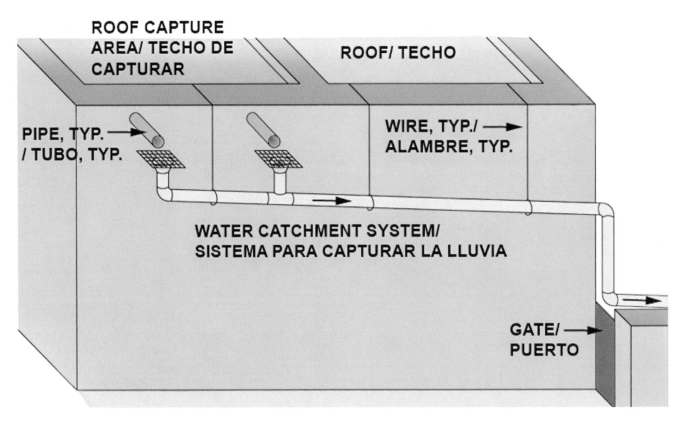

ROOF CAPTURE AREA/ TECHO DE CAPTURAR

ROOF/ TECHO

PIPE, TYP. / TUBO, TYP.

WIRE, TYP./ ALAMBRE, TYP.

WATER CATCHMENT SYSTEM/ SISTEMA PARA CAPTURAR LA LLUVIA

GATE/ PUERTO

Figure 2-3
Concrete roof to screens and piping in Parras, Mexico for use in a school garden. The concrete catchment is safe for using on plants. Only one section of the roof was used due to the small demand of the garden. appropedia.org/Daycare_rainwater_catchment_system

Figure 2-4
Clay tile roof used for a gravity fed rainwater harvesting system that supplies all the water for the main house at the Rainbow Hostel near Golfito, Costa Rica. appropedia.org/The_Rainbow_Hostel_rainwater_catchment_system

Figure 2-5
Corrugated metal roof in Santo Domingo, Dominican Republic. appropedia.org/La_Yuca_rainwater_2014

Figure 2-6
Standing seam metal roof in California, USA. appropedia.org/Campus_Center_for_Appropriate_Technology_(CCAT)

2.2 Conveyance (Gutters and Downspouts)

Conveyance is what brings the water caught on the catchment area to where it will be stored and used. Gutters, downspouts, and piping serve to convey water from the catchment area to storage and end use. Typical conveyance materials include the following:

- Bamboo (Figure 2-7).
  - Pros: Very inexpensive and easy to grow in many climates.
  - Cons: Difficult to create gutters (somewhat easier for conveyance). Lower durability.

- Hand shaped galvanized sheet metal (Figure 2-8).
  - Pros: Very accessible.
  - Cons: Hard to make joints for long lengths.

- Straight PVC (Figure 2-9).
  - Pros: Accessible and easy to work with.
  - Cons: Components can be expensive. Sizing must be more accurate.

- PVC cut in half (Figure 2-10).
  - Pros: Accessible and easy to add slope.
  - Cons: Hard to mount.

- PVC sliced and capped onto corrugated metal roof (Figure 2-11).
  - Pros: Accessible and easy to mount.
  - Cons: Hard to make necessary slope.

- Conventional "K" style aluminum, steel, zinc, vinyl, or copper. Typically, 5" or 6" (Figure 2-12).
  - Pros: Easy to mount and find fittings.
  - Cons: More expensive.

Figure 2-7

Concrete catchment with bamboo conveyance on an earthship in Haiti. appropedia.org/Haiti_Communitere_earthship_living_space

Figure 2-8
Galvanized sheet metal being hand-shaped (lower right) for gutters (left and upper right). appropedia.org/Practivistas_Chiapas_rainwater_catchment

Overflow · Flex-fitting · Zip-tie · 2. Conveyance · 3. Screen · 5. Storage · 4. First Flush · 7. End use

Figure 2-9

PVC conveyance at Zane Middle School in California, USA. The flexible fitting serves a unique innovative purpose. A zip-tie connects the screen and pipe to the awning, so that students cannot access the roof via the pipe. If a student pulls up on the pipe, the zip-tie will break instead of the pipe or awning. This has been replaced a few times (at $0.05 per fix) since installation. appropedia.org/Zane_Middle_School_rainwater_catchment

Figure 2-10
Cut-in-half PVC with a tile roof at Pedregal Permaculture Demonstration Center in San Andres Huayapam, Mexico. appropedia.org/Rainwater_catchment_at_Pedregal

Figure 2-11

PVC sliced open and pressed over the edge of corrugated metal roofing (left) in Santo Domingo, Dominican Republic. This style was eventually replaced with a conventional "K" type gutter with a splash guard (right) due to lack of slope causing over-splash, clogging, and leaking. appropedia.org/La_Yuca_rainwater_2014

Figure 2-12
"K" type PVC gutter and downspout at a farm in California, USA. appropedia.org/Bayside_Park_Farm_rainwater_catchment_system

Gutters and conveyance need to be sloped in the direction of water flow to prevent clogging, sagging, mosquitos, and damage from freezing. Rainwater harvesting systems should use approximate slope of ½ inch of drop for every 10 feet. One way to use less slope is to increase the size of the conveyance. Three common reasons to use less slope are aesthetic (so the end of the gutter is not too low), ease of mounting, and to keep water from a steep roof from overshooting the gutter.

2.3 Screens

Screens help keep a system clean. They do this by separating debris from the rainwater before storage, and often even before conveyance. Screens require maintenance, yet should serve to make that maintenance easier than if the debris was allowed to enter the rest of the system. Typical debris includes leaf litter and trash that has blown onto, or been thrown on top of, the roof.

Screens can be bought commercially (Figure 2-13) or made to order from available materials (Figure 2-14 through Figure 2-17).

Figure 2-13
Store-bought debris screen at downspout of gutter at a farm in California, USA. appropedia.org/Bayside_Park_Farm_rainwater_catchment_system

Figure 2-14
A screen (center) preventing debris from entering the three storage tanks (left) at The Rainbow Hostel located near Golfito, Costa Rica. appropedia.org/The_Rainbow_Hostel_rainwater_catchment_system

Figure 2-15
Screen at 60° angle to encourage self-cleaning (left) and that screen with some debris accumulated (right).at a middle school in California, USA. appropedia.org/Zane_Middle_School_rainwater_catchment

Figure 2-16
5-gallon bottle cut and covered with mesh to screen debris in Santo Domingo, Dominican Republic. appropedia.org/La_Yuca_rainwater_2014

Figure 2-17
Screen mesh connected to PVC with a zip-tie by Isla Urbana in Mexico. appropedia.org/ Rainwater_catchment_ at_Isla_Urbana

2.4 First-Flush

The first-flush diverts the first, and dirtiest, portion of rainwater away from the storage and remaining system components. Pollution collects on the roof between rains and is subsequently washed during the beginning of the next rain. Typical pollutants include bird droppings, ash from fires and fireplaces, and accumulated air pollution from car exhaust and industrial emissions. Long dry seasons, heavily polluted environment, and porous collection areas will raise the amount of pollution present on a roof before a rain. The first-flush prevents much of this pollution from entering the system.

Many systems are designed without a first-flush. In fact, without water quality testing, it can take a while to notice the impacts of not including a first-flush. While more studies need to be done, most studies show that removing the first part of the rain removes a significant portion of the pollutants.[8]

There are two primary types of first-flush: the tipping method and the floating ball method (Figure 2-18). Both techniques make the rainwater harvesting system healthier and longer lasting without significantly increasing maintenance requirements. Both techniques include a method to evacuate the water and reset the system between rains.

The floating ball method (Figure 2-19) uses fewer moving parts and is more durable than the tipping method. In a floating ball first-flush, the first portion of a rain fills the first-flush volume until the floating ball floats to the top and closes off the first-flush, allowing the remaining rain to run to the storage. A small borehole near the bottom of the first-flush allows for the first-flush to slowly evacuate after a rain. A removable cap allows for cleaning clogs and any debris that has found its way into the first-flush (Figure 2-20 through Figure 2-23).

That said, even with the floating ball technique, there are some important caveats and difficulties. A main source of issues is the small borehole:

---

8   Yujie, Q., De Gouvello, B., & Bruno, T. (2016, June). Qualitative characterization of the first-flush phenomenon in roof-harvested rainwater systems. In LID (low impact development conference) 2016.

- It easily clogs. Make sure to make the hole high enough so that any collected solids at the bottom of the first-flush do not block the hole. Using a removable cover for cleaning the first-flush is encouraged. In addition, attaching a tool to clean out the borehole encourages proper maintenance.

- The small hole may eject a high velocity stream of water, depending on the height of the water above it. Make sure to direct the stream in a way that is useful and does not encourage erosion.

- The hole must be made small enough so that water does not evacuate too quickly. Current research suggests a trickle to divert less water from rain events close together in time.

- Make sure to test the system by filling the first-flush with water.

Tipping versus float method first-flush.

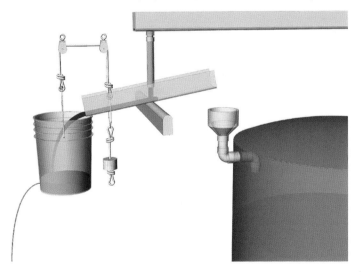

Figure 2-18
The first-flush tipping method (left) versus the float method (right). Diagrams by Gabriel Krause.

A first flush diverts the first, and dirtiest, part of the rain.

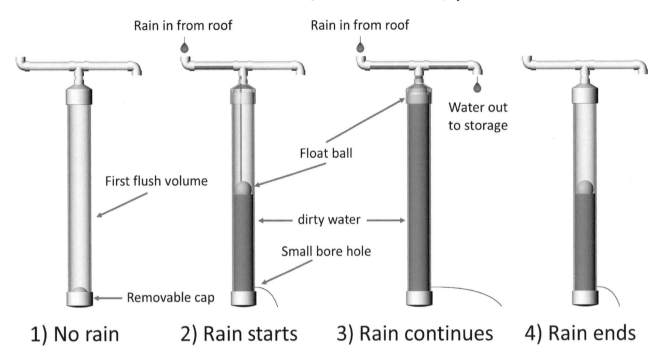

Figure 2-19

Floating ball style of first-flush. The first part of a rain fills the first-flush volume until the floating ball floats to the top and closes off the first-flush, allowing the remaining rain to run to the storage. A small borehole allows for the first-flush to evacuate after a rain. Please note that the borehole stream is exaggerated for demonstration purposes. A removable cap allows for cleaning clogs and any debris that has found its way into the first-flush. Diagram by Gabriel Krause.

Figure 2-20

First-flush (large vertical pipe on left) with the filter above and storage to the left. The removable cap is off and the float ball removed (at bottom on left). Daycare garden system in Parras, Mexico - appropedia.org/Daycare_rainwater_catchment_system. 208-liter first-flush (right) designed by Isla Urbana. appropedia.org/Rainwater_catchment_at_Isla_Urbana

Figure 2-21
Two 4" PVC pipes (left) are used, instead of just one, to make the first-flush of necessary volume, using the parts that were available in Santo Domingo, Dominican Republic. One pipe large enough was not available. Each pipe has a drain hole at the bottom pointing away from the wall. Removable caps are at the end of each T at the bottom of the first-flush - appropedia.org/La_Yuca_rainwater_2014. A 20-gallon barrel (right) at the bottom of a downspout T at the Arcata Sanctuary art non-profit in Arcata, California - appropedia.org/Sanctuary_rainwater.

Figure 2-22

A 15-gallon first-flush at the Campus Center for Appropriate Technology in California, USA. This clear downtube (left) filled with algal growth. It was eventually replaced with an opaque PVC pipe (right) to prevent algal growth. appropedia.org/CCAT_rainwater_catchment_system

Figure 2-23
A partially cocked valve can be used instead of a small borehole to slowly evacuate the diverted water. This allows for easy cleaning of the system by periodically opening the valve completely and reclosing it to a partially cocked position.

The water from the first-flush diversion can still be used for end uses that do not need to be clean, such as cleaning equipment or watering ornamental plants.

The partially cocked valve solution in Figure 2-23 is slightly more expensive than a small borehole, due to the cost of the valve as opposed to just drilling a hole. This extra cost can be offset by not needing a removable bottom cap. The borehole method does introduce one interesting social issue—people will often close the valve all the way if they see it dripping. The valve needs to continue to drip, so that the first-flush will be fully evacuated and ready for the dirty water from the next rain. Therefore, appropriate signage or a method to prevent fully closing the valve is needed.

2.5 Storage

Storage holds water for later use. Later use can be between frequent rains, requiring less storage volume, or in the dry season, requiring more storage volume. Typical storage for built (active) systems includes plastic storage tanks (Figure 2-24), commercial rainwater tanks, 55-gallon drums (Figure 2-25), custom ferrocement (cement or lime, with sand, applied over a metal mesh such as fencing) tanks (Figure 2-26), and Intermediate Bulk Containers (IBCs) (Figure 2-27).

Figure 2-24

Examples of plastic tanks. 2500 gallons (upper left) for garden and emergency use in Eureka, California, USA. appropedia.org/M_Street_Eureka_ rainwater_catchment. 1500 gallons (upper right) for household use in Oregon, USA. appropedia.org/Ersson_rainwater_harvest_and_purification. 500 gallon (bottom) for school garden use in Eureka, California, USA. appropedia.org/Zane_Middle_School_rainwater_catchment

Figure 2-25
A 900-gallon pond lined hole (left) in a wet system where the pipe passes under a path and back up into the 55-gallon drum at a community center in Arcata, California, USA. appropedia.org/Sanctuary_rainwater. The 2,000 liter 'rainjar' in use in Northeast Thailand (right) can cost as little as $20 USD. appropedia.org/Rainwater_harvesting

Figure 2-26
Ferrocement tanks of various sizes: A 19,000-liter tank (left) for use by a small community in Chiapas, Mexico. One incorporated into an earthship structure (center) at Haiti Communitaire. A 9,950-liter tank (right) with a clay plaster at a Permaculture Demonstration Center in San Andres Huayapam, Mexico.

Figure 2-27

275-gallon Intermediate Bulk Containers (IBC) being prepared for water storage aboard Swale, a floating food forest in New York. See appropedia.org/Working_with_IBC_totes for more on working with IBCs. appropedia.org/Swaleny

In all systems, it is important to think about weight, pest avoidance, and vacuum prevention. A vacuum can be introduced in a relatively sealed system by quickly evacuating the storage. Once a vacuum is introduced it can prevent the system from continuing to function. You can create an example vacuum by (1) filling a bottle with water, (2) inverting the bottle under water, (3) raising the bottle while keeping the lip below the waterline, and (4) observing that the jar is still full of water even though it is mostly above the waterline (Figure 2-28). This water would evacuate if you drilled a small hole in the base of the bottle, or if you raised the bottle above the water line, allowing air in.

vacuum

water in bottle suspended above river

lip below river

Figure 2-28
Demonstrating a vacuum with a bottle of water suspended above the water line of a river.

To prevent any vacuum, a breather hole (a hole above the water line that allows air to flow freely in and out of the system) should be introduced, and is usually included in any conventional plastic tank. This breather hole can be double purposed with the overflow pipe if necessary (Figure 2-29). In both cases, any holes should be covered with mosquito mesh to prevent mosquitos (in applicable regions). The weight of the tank can be calculated using the formulae in Section 3.1. The weight can be critically important, as systems can fail if the storage becomes too heavy for the platform, or worse, the roof or hill on which it stands (Figure 2-30).

Figure 2-29
Overflow and breather hole covered with mosquito screen in Santo Domingo . appropedia.org/La_Yuca_rainwater_2014

Pipe bending between barrels due to broken pallet

60-gal pickle barrel too heavy for uncompacted earth and pallet

Figure 2-30
The weight of the full 60-gallon pickle barrel broke the pallet and displaced the uncompacted earth below the pallet . This breakage resulted in bent pipes and junctions, threatening the function of the entire rainwater harvesting system.

2.6 Water Purification

Water purification further cleans rainwater, depending on the end use and need for purity. Rainwater harvesting water, especially in systems that use a first-flush, can be used for many applications without purification. For example, watering ornamentals needs no first-flush or purification, and watering edible plants depends on context. However, potable water always requires some purification. Some typical examples of that purification (also known as treatment) include:[9]

- **Canister filters** most often use conventional sedimentation, ceramic, and activated carbon filters (Figure 2-31).
  - Pros: Easy to incorporate, precise.
  - Cons: More expensive.

- **Activated carbon filters** use chemical adsorption to attract contaminants to its surface. Can be incorporated into a canister filter.
  - Pros: Removes volatile organic compounds, tastes and odors. Removes chlorine. Works great as part of a treatment chain with other filters.
  - Cons: Not effective at removing viruses and bacteria.

- **Ceramic filters** use a labyrinth of microscopic holes to filter out contaminants based on size.
  - Pros: Locally manufacturable. Cleanable with a brush.
  - Cons: Slow flow rates. Can clog easily. Usually not effective against viruses, especially if made locally.

---

9 Water purification could be its own book. See more at http://www.samsamwater.com/library/RAIN_ Rainwater_Quality_Policy_and_Guidelines_2008_v1.pdf   and   http://www.appropedia.org/Water_ purification.

- **Hollow fiber membrane filters** use a semi-permeable barrier to filter out contaminants based on size. Often incorporated into a canister.
  - Pros: Small micron filters (0.1 micron) can effectively filter out bacteria, protozoa, or cysts, even smaller (0.02 micron) can also filter out viruses. Long life and relatively high flow rates compared to a ceramic filter.
  - Cons: Can clog easily when used with dirty (high sediment) water. Relatively expensive upfront cost. Bad tastes and odors are not removed like they are with activated carbon.

- **Slow-sand filters** (Figure 2-31 and Figure 2-32) use layers of sand and gravel with a developed biological layer on top called a Schmutzdecke.
  - Pros: Inexpensive and accessible.
  - Cons: Slow; large area needed; less precise.

- **Bioremediation** uses living organisms as treatment, such as mycoremediation with fungi or phytoremediation with living plants.
  - Pros: Resilient and regenerative.
  - Cons: Less precise[10] and more sensitive.

- **SoDis** (Solar water Disinfection) uses plastic bottles in sunlight to purify water using heat and UV.
  - Pros: Very inexpensive and accessible.
  - Cons: Only does disinfection, not purification. Needs the sun and low turbidity water (which is easy to achieve with rainwater).

- **Solar pasteurization** uses the sun to bring water to a certain temperature, below boiling, for a certain amount of time.
  - Pros: Lower energy requirements than boiling.
  - Cons: Only does disinfection, not purification. Needs the sun and larger area.

- **Solar distillation** uses the sun to hasten the evaporation of water and collects the condensation on a surface.
  - Pros: Relatively inexpensive energy compared to other fuels. Produces cleaner water than some other methods, i.e. does purification.
  - Cons: Large area needed; more expensive. Relies on the sun.

---

10 Bioremediation is much easier to incorporate into landscape type rainwater catchment, as opposed to a built type.

- **Boiling** uses energy, usually from wood or fossil fuels, to boil the water.
  - ○ Pros: Accessible and common.
  - ○ Cons: Expensive in energy and time. Only does disinfection, not purification.

- **Reverse Osmosis** uses electrical energy to apply pressure and push water through a water-permeable membrane.
  - ○ Pros: Can be used with brackish water. Produces reliably clean water (except when filter is compromised physically or biologically).
  - ○ Cons: Very expensive and sensitive. Energy intensive. Must dispose of brine waste.

- **UV** uses ultraviolet radiation, usually from an electrical lamp, to prevent microorganisms from reproducing by altering their DNA.
  - ○ Pros: Very effective against viruses and bacteria, as long as the water is visually clear.
  - ○ Cons: Expensive and energy intensive. Does not work on non-biological contaminants.

- **Chlorination** uses any one of a number of forms of chlorine to disinfect water by killing many pathogenic (disease-causing) organisms.
  - ○ Pros: Very common for water disinfection and effective at treating many of the most prevalent forms of waterborne illness, e.g. cholera, typhoid, and dysentery.
  - ○ Cons: Needs careful management as chlorine and its products can be very toxic. Needs a supply of chlorine.

- **Electrochlorination** uses an electric current through salt water to produce sodium hypochlorite. This is just a special form of chlorination, but deserves special mention for the following pros that make it a strong candidate for village-scale potable water:
  - ○ Pros: Common and effective for water disinfection and treating many of the most common forms of waterborne illness. Does not need a supply of chlorine. Produces chlorine at relatively safe levels.
  - ○ Cons: Requires an electrical energy source. Requires mechanical and electrical expertise.

Figure 2-31
The slow sand filter (left) was not trusted by the community (see Section 5.2 for more on this) and eventually was replaced by canister sediment and carbon filters (right). appropedia.org/La_Yuca_rainwater_catchment_2013

Figure 2-32

Samples from before and after a slow sand filter from Swale, the floating food forest in New York. The pre-filter water (left) is rainwater that has infiltrated into the bilge (compartment under the deck) of the barge, where it becomes much dirtier. The post-filter water (right) is taken directly after slow-sand filtration. The clearer water on the right indicates efficacy of the slow-sand filter. Further laboratory testing proved it as well. Credit: Liz Lund, CC BY-SA

It is possible and often desirable to use the above listed methods of water purification in series for extra treatment. For example, systems for use as drinking water may use a slow-sand filter followed by UV treatment. The slow-sand filter cleans the water so that a relatively smaller UV lamp can be used for final purification of biological activity.[11]

Different treatment methods treat different pathogens. For example, UV is more effective against parasites whereas chlorine is more effective against viruses.[12]

2.7 End Use

An end use refers to how and where the water eventually will be used. There is no great reason to catch rainwater without an end use. Typical end uses include water for drinking, gardening, landscaping, cleaning, etc.[13] As mentioned before, potable water is a greater challenge than other end uses, which may include cleaning, flushing, washing, watering, etc.

In addition to the level of purity needed, the amount of pressure needed for end use will also affect your choices. Some systems need significant pressure, e.g. showering, where as other systems need less, e.g. drip irrigation. See the images in Figure 2-33 and throughout this book for more examples of different end use cases. The larger the

---

11 Most treatment methods are not effective at removing heavy metals. A few specialized activated carbon filters can remove heavy metals. Most rainwater catchment systems do not have dangerous levels of heavy metals. Heavy metals can be introduced from industrial dust and pollution, or from the construction materials for the rainwater catchment system. If in doubt, you may want to test for heavy metals.

12 Texas Commission on Environmental Quality (2007). Harvesting, Storing, and Treating Rainwater for Domestic Indoor Use. Retrieved from http://rainwaterharvesting.tamu.edu/files/2011/05/gi-366_2021994.pdf

13 Many other end uses exist. I once assisted on a system for pre-soaking wine barrels before they are used for wine. That was a tasty and fruitful project.

difference in vertical height between water storage and water outlet, the greater the pressure, as discussed in Section 3.

Figure 2-33
This system in California, USA provides water storage for emergencies, such as an earthquake, and for use in the gardens and chicken coop. appropedia.org/M_Street_Eureka_rainwater_catchment

2.8 Labels and Signs

All systems need labels or signage that instruct and direct users. The most important label shows whether the system is potable or not. In addition to labels, consider nudges and point-positive design in your system.

Nudges help users to do the right thing by gently reminding them. For example, if a user is intended to clean the borehole regularly, consider placing the first-flush nearby and facing a commonly used path so that the user sees the hole and is reminded.

Point positive design focuses on what a user should do, instead of what they shouldn't do. For example, if a system is for use only on ornamental plants, in addition to the "Not

Potable" sign, consider placing permanent irrigation in a way that reaches only the ornamentals and not the edibles or a drinking location.

Communication through signs is a subtle art. Consider using common symbols and as few signs/words as possible to avoid over-saturation. In addition, prototype your signage. Prototyping is a way to test early whether your signage works. Prototype signs can be accomplished with cardboard and markers, before upgrading to more indelible materials. When prototyping, watch how people use the system without your guidance to see what errors might arise. The confusing signage at the Bayside Park Farm, a community supported agriculture and educational park in Bayside, California, resulted in rainwater being significantly underused. The sign (Figure 2-34) stated that the water from the first-flush should not be used on plants. After some turnover of employees, that sign was interpreted to mean that none of the rainwater (not just the first-flush water) should be used on crops. Therefore, the water was only being used for tool cleaning, which in turn meant that there was much more water storage than necessary.

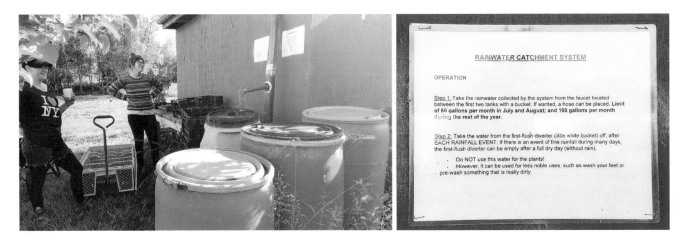

Figure 2-34
Confusing signage at the Bayside Park Farm in Bayside, California . New staff understood the signs, which stated, "Do NOT use this water for the plants!" to mean that none of the collected rainwater was usable for plants—when in fact the warning was intended to apply only to first-flush water.

Since safety is paramount, all non-potable rainwater systems, especially those accessible to the public, should be well labeled to prevent drinking. Tanks or spigots of untreated water should be labeled according to local law or custom to prevent accidental consumption (Figure 2-35).

Figure 2-35
Template warning sign for all untreated rainwater.

3 Pressure

Pressure is critical to moving water from where it is gathered or stored to where it will be used. In a rainwater harvesting system, the water must be able to flow from catchment through the filters and conveyance into the first-flush and storage or end use. This pressure can be provided by gravity from vertical height difference or by a pump.

As an idea of how much pressure you will need: typical US residential water pressure is between 40 to 80 psi (pounds per square inch); typical drip irrigation systems (and some micro-sprinklers) need between 15 to 25 psi; and some appropriate technology drip irrigation systems need only 4 to 10 psi.

In addition to flow from catchment to storage, the flow from storage to end use is critical. Using the existing topography and/or platforms can often yield enough pressure for end use. If necessary, a pump can be implemented to add sufficient pressure. While utilizing a pump increases the pressure, it also increases the upfront and operational costs.

Gravity acting on the vertical height of the water column is what produces the pressure, which is also referred to as *head*. Make sure not to confuse volume with pressure (head). For instance, a 20-foot-tall water tower of 8000 gallons has the same water pressure as a 20-foot-tall pipe of 80 gallons. To illustrate this effect, picture (or build) a U-pipe with two different diameter legs and a valve in between them (Figure 3-1). Then fill the larger diameter pipe with water. Then see where the water level will be once you open the valve. Since water pressure is independent of volume, the levels should be equal to each other (also referred to as "water seeks its own level").

On the left, the thin pipe is empty. Consider what will happen to the water levels when the red valve is opened.

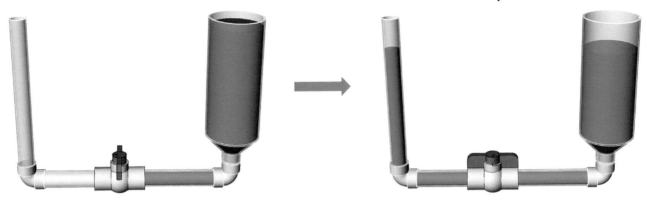

Water seeks its own level when connected (and exposed to same pressure).

Figure 3-1

A U-pipe with two different diameter pipes (left) starts with a closed gate valve to keep the larger diameter pipe full. When the valve is opened (right) and water can flow between pipes, notice that the water level settles at the same height in both pipes. Water pressure is independent of volume; it is a function of depth, not breadth. Diagram by Gabriel Krause.

Another experiment to convince yourself of water pressure being dependent on height not volume is to feel the pressure at the bottom of a 10-foot-deep pool compared to 10 feet deep in a freshwater lake. Those two pressures should feel the same. Therefore, the head available in a system can be determined by measuring the available heights and applying basic physics or a standard conversion factor.

3.1 Mass and weight

Before we calculate pressure, let's calculate the mass of our storage. As stated before, weight is critically important as systems can fail if the storage becomes too heavy for the platform, or worse, too heavy for the roof or hill on which the storage sits. In addition, knowing how to calculate the weight of storage leads well into calculating pressure, which is force over area where the force is the weight of the water.

To calculate mass, we use the following formula:

Equation 1
Mass from density and volume.

$$m = \rho * V$$

Where:

- m = mass

- ρ = fluid density (i.e. mass/volume, which for water is approximately 1,000 kg/m^3 or 62.4 lb/ft^3 or 8.34 lb/gal)[14]

- V = volume of storage
  For the 500-gallon tank in Figure 2-24, calculate the mass in pounds:[15]

---

14 This is at 4°C. That said, we round most things in this book to around three significant figures, so this is close enough to true for all our rainwater environments. Common conversions in Section 6.1 - Units Related to Water.

15 On Earth, lbm (pounds mass) and lbf (pounds force) are equal. We will refer to them both as lb.

$$m = \rho * V = 8.34\frac{lb}{gal} * 500 \; gal = \textbf{4,170 } \textbf{\textit{lb}}$$

Knowing that 1 ton = 2,000 lb, we can see that that 500-gallon tank is:

$$4,170 \; lb * \frac{1 \; ton}{2000 \; lb} = \textbf{2.085 } \textbf{\textit{tons}}$$

So, our 500-gallon tank weighs just over 2 tons, which is around the same weight as a small car, three dairy cows, or over 20 adult humans!

For the 19,000-liter ferrocement tank in Figure 2-26, calculate the mass in kg, remembering that in SI units (the International System of Units, e.g. meter, kilogram, second, etc.) 1,000 liters equals 1 m^3 so:

$$19,000 \; liters \cdot \frac{1 \; m^3}{1,000 \; liters} = \textbf{19 } \textbf{\textit{m}}^{\textbf{3}}$$

And:

$$m = \rho * V = 1000\frac{kg}{m^3} \cdot 19 \; m^3 = \textbf{19,000 } \textbf{\textit{kg}}$$

At this point you might notice how awesome SI units really are, but they are even more awesome than that; remember that 1 kilogram is defined to equal 1 liter of water, so:

$$m = \rho * V = 1\frac{kg}{liter} * 19,000 \; liters = \textbf{19,000 } \textbf{\textit{kg}}$$

Note that this is quite heavy and is approximately the weight of 10 Tesla Model S60 cars or approximately 300 adults.

3.2 Calculating Pressure

Pressure is critical because it is what moves water from one point to another. A common constraint when designing a rainwater harvesting system is providing enough pressure for the end use. For systems with low roofs or challenging topography (e.g. the catchment area is in a depression), sometimes sufficient pressure can only be obtained with a pump. Pressure from the vertical height of water, also called *head*, can be found from the following equation:

Equation 2

Defining pressure as force over area

$$Pressure = \frac{Force}{Area}$$

The weight density of water[16] is 62.4 lb/ft³, which is force/volume. Knowing that Volume is equal to Area * height (V=A*h), we can convert that weight density into head by multiplying it by the vertical height of the water.

$$Water\ Pressure(P) = \frac{Force}{Area} = 62.4\frac{lb}{ft^3} * height\ of\ water$$

As an example, find the water pressure from water that is 1 foot high:

$$P = 62.4\frac{lb}{ft^{3^2}} * 1ft = 62.4\frac{lb}{ft^2}$$

Therefore, the pressure from 1 foot of water is 62.4 lb/ft². Unfortunately, you won't usually find pressure measured in these units. To convert lb/ft² into the more conventional psi (pounds per square inch), use the fact that 1 foot is equal to 12 inches:

---

16 This is the weight per volume and specifically for water around 50°F. The value changes to 62.43 lb/ft³ at 40°F. http://www.engineeringtoolbox.com/water-specific-volume-weight-d_661.html

$$P = 62.4\frac{lb}{ft^2} * \frac{1ft}{12\ in} * \frac{1ft}{12\ in} = 0.433\frac{lb}{in^2} = 0.433\ psi$$

This shows that the pressure from 1 foot of water is 0.433 psi, which is the basis of the commonly used conversion:

Equation 3
Field conversion for head.

0.433 psi for every vertical foot of water

You can use either process to find the pressure that is due to the vertical height of water. For example, to solve for the pressure exerted by 20 feet of water (e.g., at the bottom of a full 20-ft-tall tank):

$$P = 62.4\frac{lb}{ft^{32}} * 20ft = 1,248\frac{lb}{ft^2} * \frac{1ft}{12\ in} * \frac{1ft}{12\ in} = 8.67\frac{lb}{in^2} = 8.67\ psi$$

Or

$$P = \frac{0.433\ psi}{ft\ of\ water} * 20\ ft = 8.66\ psi$$

Therefore, 20 vertical feet of water exerts a pressure of 8.66psi. That pressure is sufficient for hand washing, watering, and most drip irrigation lines.[17]

In addition, keep in mind the differences between static pressure and dynamic pressure. Static pressure refers to the pressure when the water is not flowing. Dynamic pressure refers to the pressure when the water is flowing. Static pressure is always higher than dynamic pressure because friction on the moving water reduces the pressure available at the bottom of the system. If you determine the static pressure to be just enough to get from one point to another, there actually might not be enough to get there because of dynamic pressure loss when the water is moving. The longer the pipe and the smaller the diameter, the more loss there will be.

---

17 The small (0.01 psi) difference is due to rounding error. Also keep in mind that this is static pressure, so we will have less dynamic pressure actually available (due to pressure loss from moving water).

4. Calculations

The previous section on gravity and the following sections on calculations are used to size the necessary components of built (i.e. active) rainwater harvesting systems like the ones this book covers. These components can often be calculated with rules of thumb, but the more in-depth calculations will engender a deeper understanding and the ability to adapt to more customized systems. Laboring through the calculations builds your toolbox and ability to apply knowledge between different types of rainwater systems, and even between entirely different systems such as rainwater and greywater.

4.1 Usage

Rainwater harvesting systems are typically sized based upon supply or demand. In either case, it is important to calculate, and maybe even conserve, demand.

Many sources of demand can exist for a system, with varying methods to determine their relative demands, e.g.:

- Sinks
- Showers
- Dishwashing machines

- Clothes washing machines
- Toilets
- Garden hoses

- Farm irrigation
- Leaks
- Livestock

Washing machines and toilets are measured in volume per use. Here are some typical values:

- Dishwashing machines: Many older machines are 10-15 gal/cycle. Many Energy Star machines use 4-6 gal/cycle.

- Clothes washing machines: Many older machines are 40-45 gal/cycle. A full-size Energy Star machine uses approximately 13 gal/cycle.

- Toilets: Many older toilets are 3.5-7 gal/flush (GPF). A Watersense High Efficiency Toilet uses just 1.28 GPF.

After determining the water usage of your washing and flushing machines, you can count or estimate the number of usages per a certain length of time (e.g. per month) to determine your demand from those machines.

Sinks, showers, garden hoses, and various household leaks, among other items, are determined by their volumetric flow and time used. Volumetric flow rate "Q" is the volume of water passing through a cross-sectional area in a given amount of time. Two common examples are gallons/minute (GPM) and liters per second (LPS). Volumetric flow is defined in the following formula:

Equation 4
Volumetric flow.

$$Q = \frac{V}{t}$$

- V = volume
- Q = volumetric flow
- t = time

Therefore, volume can be written as:

$$V = Q * t$$

For example, a low flow showerhead is 1.5 GPM. If you shower for a leisurely 11 minutes, the volume is:

$$V = Q * t = 1.5 \frac{gal}{\cancel{min}} * 11 \frac{\cancel{min}}{use} = 16.5 \frac{gal}{use}$$

If you do that daily:

$$16.5 \frac{gal}{\cancel{use}} * 1 \frac{\cancel{use}}{\cancel{day}} * 30 \frac{\cancel{days}}{month} = 495 \frac{gal}{month}$$

Farm irrigation and livestock demands are dependent on many environmental and technical factors. Determining those demands are outside the scope of this book. Looking at existing use for each month in the past year (or even better, longer term monthly averages) at your location is a great place to start.

4.2 Catchment Area

Rain falls. In rainwater catchment, our job is to catch that rain and use it with purpose. The catchment area is the area which intercepts the rain. Knowing the catchment area is the first step in calculating the volume of water that can be caught from rainfall. A flat roof is a great example of a catchment area. Another example is an open barrel sitting alone out in the rain, in which the catchment area would just be the circular open top of the barrel.

To determine a catchment area, assume that the rain falls straight down.[18] That means that no matter what the pitch of your roof is, it is just the footprint (i.e. the vertical projection) that determines how much rain you can catch (i.e. catchment volume).

The two buildings in Figure 4-1 have identical floor plans and different roofs. The flat roof (left building) is equal to both the footprint and the collection area. The pitched roof (right building) has a larger roof area than the flat roof. Imagine laying the pitched roof flat; its area would be much greater. If the roof area on the left is 1200 square feet, the roof area on the right is 1600 square feet. However, since the rain is falling straight down, it only sees the 1200-square-foot collection area created by the identical 1200-square-foot footprint of either building.

---

18 Although this isn't always the case. I once worked in a community near a volcano in El Salvador where the rain always seems to fall at a steep and specific angle that was not straight down.

Rain collection area is based on horizontal area.

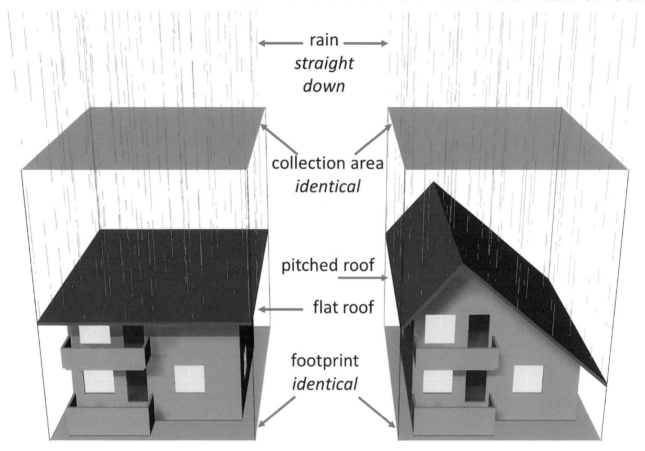

Figure 4-1

Two houses with the same footprint and different roofs. The flat roof (left) has a smaller roof area than the sloped roof (right). However, the footprints (bottom square) are identical, so the catchment areas (top) are also identical.

For a building with a roof of 1400 ft², and a footprint of 60 ft x 20 ft, the catchment area is calculated to be 1200 ft² using the area of a rectangle represented by the following equation:

Equation 5
Area of a rectangle.

$$A_{rectangle} = length * width$$

$$A = 60 \; ft * 20 \; ft = 1200 \; ft^2$$

4.3 Gutter and Downspout Sizing

Once the rain falls on the roof, gutters (i.e., conveyance) are usually needed to direct the rainfall to any treatment, storage, and/or end use. Pipes that are too small will restrict water from flowing through the system fast enough, resulting in overflow or overloading. Pipes that are too large will convey the water easily, but could prove cost-restrictive or unsightly. More slope will help evacuate the pipes faster, but may be harder to build and less attractive.

There are many ways to size gutters. One of the simplest is the following rule of thumb:[19]

Equation 6
Gutter sizing rule of thumb based on area.

$$\frac{1 \; cm^2 \; of \; gutter \; cross \; section}{1 \; m^2 \; of \; roof \; area}$$

Four other methods are:

---

19 More detailed gutter sizing information at http://www.appropedia.org/Rain_gutter_sizing

- If you are using a standard gutter, e.g. "K" style gutters, many online calculators exist.

- If you are using standard pipe, pipe friction tables can be used to find an acceptable amount of pressure (head) loss.[20]

- For full pipes of various materials and sizes, the Darcy-Weisbach method (or easier, the Hazen-Williams method) can be used to find an acceptable amount of head loss.

- For open channel flow, the Manning Equation can be used to find an acceptable amount of head loss.

The four other methods are more accurate than the simple rule of thumb, and some allow for factors such as roof pitch, strength of local storms, and other shapes besides cylindrical pipe.

Example

The minimum pipe size needed for a catchment area of 23m^2 is calculated to be 23 cm^2 using the rule of thumb in Equation 6 as shown below:

$$gutter\ cross\ sectional\ area = 23\ m^2 * \frac{1\ cm^2}{1\ m^2} = \boldsymbol{23\ cm^2}$$

The needed cross-sectional area of gutter is 23 cm^2. To convert that cross-sectional area to diameter of pipe, remember that the formula for the area of a circle (which is the cross-section of a pipe) is the following:

---

20 The pressure (head) loss methods (friction tables, Darcy-Weisbach, Hazen-Williams, and Manning) are covered at http://www.appropedia.org/Rain_gutter_sizing.

Equation 7
Area of a circle

$$A_{circle} = \pi * r^2$$

Where:

- A = Area

- $\pi \approx 3.1416$

- r = radius, which is half of the diameter (d/2).

Therefore:

$$A_{circle} = \pi * \left(\frac{d}{2}\right)^2$$

Solving for diameter yields:

$$d = 2 * \sqrt{\frac{A_{circle}}{\pi}} \Rightarrow d = 2 * \sqrt{\frac{23\,cm^2}{\pi}} = 5.41\,cm$$

Converting the diameter to inches (which is the common unit of measure for pipe size) yields:

$$5.41\,cm * \frac{1\,in}{2.54\,cm} = 2.13\,in$$

Therefore, a pipe diameter of at least 2.13 inches should be used. The most common size that meets that requirement is 2.5-inch pipe.

This rule of thumb works for most situations, but ultimately the sizing depends on the strength of local storms,[21] the pitch of the roof, and the slope of the gutter.

In addition to gutter sizing, gutter slope must be addressed. In order to keep the water flowing through gutters via gravity, a slope is needed to provide pressure from the elevation difference (Figure 4-2).

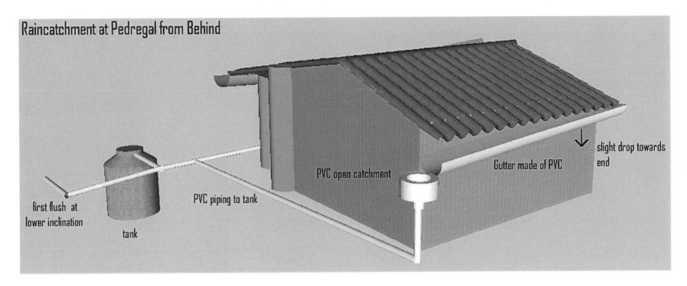

Raincatchment at Pedregal from Behind

slight drop towards end

Gutter made of PVC

PVC open catchment

PVC piping to tank

first flush at lower inclination

tank

Figure 4-2
Gutters showing an exaggerated drop in the direction of water flow at Pedregal, a Permaculture Demonstration Center in San Andres Huayapam, Mexico. appropedia.org/Rainwater_catchment_at_Pedregal.

A common ratio for the needed slope is 1/2 inch drop for every 10 feet of run, as shown in Equation 8.[22]

---

21   The strength of local storms is often found as the maximum 60-minute-long storm event, 100-year return period, in a location - Kniffen, B., Clayton, B., Kingman, D., Jaber, F. (2012). Rainwater Harvesting: System Planning. Fort Stockton, TX. Texas A&M University. Pg 71.

22   https://www.thisoldhouse.com/how-to/how-to-install-rain-gutters. Also see slopes and roof drainage - http://www.engineeringtoolbox.com/sloopes-roof-drainage-d_1107.html

Equation 8
Gutter slope ratio.

$$\frac{\frac{1}{2} \ in \ of \ drop}{10 \ feet \ of \ gutter \ run}$$

Using the common gutter slope ratio, calculate the drop needed for a 30-foot-long gutter.

$$drop = 30 \ ft * \frac{\frac{1}{2} \ in}{10 \ ft} = 1.5 \ in$$

Therefore, one end of the 30-foot gutter should be approximately 1.5 inches above the other to ensure proper flow through the gutters.

4.4 First-Flush Sizing

There is no single true method to determine the first-flush volume. This lack of assuredness is due to the great environmental variability of systems, including the level of pollution present, the ease of washing away pollution from roofing material, the time between rains, the strength of the rain, etc. As a rule of thumb, contamination is halved for each mm of rainfall flushed away.[23] Following are two ways—area based and exponential decay based—to determine the appropriate size of a first-flush. Both methods contain assumptions. The exponential decay based model contains fewer assumptions. For a metal roof in a suburban neighborhood, the area based rule of thumb should suffice. For a clay tile roof on a dirt road, especially for water needed for drinking, the exponential decay model may be necessary.

---

23 Martinson, B., & Thomas, T. (2005). Quantifying the first flush phenomenon. In 12th International Rainwater Catchment Systems Conference.

A final consideration is the length of time between reset of the first-flush and the next rain. A reset refers to the first-flush going from full to empty, and therefore ready to catch the next rainfall. Ideally all the first rain after a long dry season would be diverted from storage and discharged to untreated usage. Then, the first-flush would slowly evacuate so that subsequent rains would have only some of the initial water diverted. These aspects still need more study globally, in order to determine the best practices for the many different combinations of locations, roof materials, pollutants, etc.

Area-based rule of thumb:

The area-based rule of thumb assumes a roof that is easily cleaned and in a clean environment. It is a simple and most commonly used rule of thumb, yet its effectiveness is currently being debated in literature and practice. The area-based rule of thumb is fast and works well in many settings such as a relatively clean environment and non-porous roof (such as metal). The following two formulae represent the area-based rule of thumb in imperial and SI units:

Equation 9
First-flush volume area-based rule of thumb in Imperial units.

$$first\text{-}flush\ volume = \frac{1\ gal}{100\ ft^2\ of\ roof}$$

Equation 10
First-flush volume area-based rule of thumb in SI units.

$$first\text{-}flush\ volume = \frac{0.41\ liters}{m^2\ of\ roof}$$

Exponential Decay Model:

The exponential decay model is more explicit in its assumptions than the area-based rule of thumb. The model requires a decay value that has only been experimentally found for a few roofs, rains, and environments. In addition, the exponential decay model requires testing water turbidity (the cloudiness of the water).[24] The exponential decay model is represented in the following formula:

Equation 11
Exponential decay based model (SI units) for first-flush precipitation.

$$V_{ff} = -\frac{\ln\left(\dfrac{target\ turbidity}{runoff\ turbidity}\right)}{\lambda} * A * k$$

Where:

● V_{ff} = Needed volume of first-flush in liters

● ln = is the natural log function (it is present on most calculators and in Excel)

● Target turbidity = the target turbidity entering the tank in Nephelometric Turbidity Units (NTU).
  ○ The World Health Organization states a target of 5 NTU for water leaving the storage tank (i.e. after storage and before use).[25]
  ○ 20 NTU entering the tank is usually sufficient. [26]

● Runoff turbidity = the measured average runoff turbidity from the catchment area.
  ○ A somewhat dirty roof might contribute just 20-100 NTU.
  ○ A very dirty roof might contribute 1000 NTU. [26]

---

24 Measuring turbidity can be done at a lab and many universities. Also Appropedia documents an open source design to measure turbidity at http://www.appropedia.org/Open-source_mobile_water_quality_testing_platform.

25 Cobbina, S. J., et al. "Rainwater quality assessment in the Tamale municipality." Int. J. Sci. Technol. Res 2 (2013): 1-10.

26 Martinson, B., & Thomas, T. (2005). Quantifying the first flush phenomenon. In 12th International Rainwater Catchment Systems Conference.

- λ = the exponential decay value.
  - ◊ These values are experimentally found.
  - ◊ A very clean roof may be as high as 2.2/mm. A very dirty roof may be as low as 0.7/mm.

- A = catchment area in square meters

- k = conversion factor to convert from mm*m² into liters. That conversion is 1.

Finally, for potable use, filtration should be used to bring the turbidity to below 1 NTU[27] depending on any additional treatment.

4.5 Catchment Volume

The catchment volume is calculated from the precipitation falling on the collection area with some loss due to the efficiency of the collection materials (and leaks). In addition, conversion factors are used to yield the desired units of volume. Typically, monthly catchment values are calculated based on monthly average precipitation data. The collection volume for any period of time is calculated using the following formula (mnemonic device Vrake):

Equation 12

Vrake - Rainwater harvesting potential.

$$V = R * A * k * e$$

---

27 National Primary Drinking Water Regulations. (2017, March 21). Retrieved from
https://www.epa.gov/ground-water-and-drinking-water/national-primary-drinking-water-regulations.

Where:

- V = Volume of collection in gal/time or m³/time or liters/time
  - Note that time is usually in months.
  - Use this to help determine potential yield and tank size.

- R = Precipitation in inches/time or mm/time
  - Collect this data or find it from existing climate data.

- A = Footprint of collection surface in ft² or m²
  - This is the vertical projected area of the collection surface. For a rectangular house, use length times width.

- k = Needed conversion factors, such as 7.48 gal/ft³ or easier SI units
  - Can also combine the 1ft/12in conversion for the precipitation data here.

- e = Efficiency of collection surface (which is unitless)
  - 0.75 soil, 0.8 average, 0.95 metal[28, 29]

Imperial Units Example

Calculate the collection potential (in gallons) for a 1,900 square foot single-story house, with a slanted shingle roof, in Columbia, Missouri for the month of March.

The first step is finding the precipitation data for your location. For example, the table in Figure 4-3 shows the precipitation data for Columbia, Missouri.

28 Haan, C. T., Barfield, B. J., & Hayes, J. C. (1999). Design Hydrology and Sedimentology for Small Catchments. San Diego u.a.: Academic Press.

29 Waterfall, P. (2006). Harvesting Water for Landscape Use. Retrieved from https://extension.arizona.edu/sites/extension.arizona.edu/files/pubs/az1344.pdf

1971 to 2000 Rainfall Averages (inches) for Columbia, Missouri

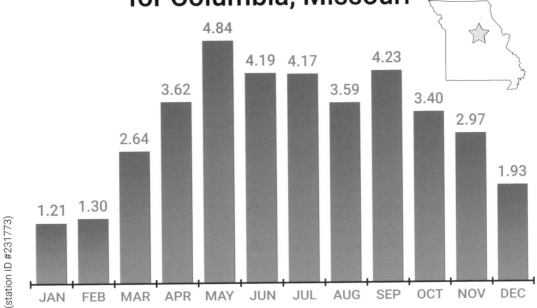

Figure 4-3

Average monthly precipitation for Columbia, Missouri. Graphed from NOAA and ggweather.

Looking up the efficiency of the collection surface for a shingle roof yields an average 0.8.

$$V = R * A * k * e$$

$$V_{March} = 2.64 \frac{in}{March} * \frac{1\,ft}{12\,in} * 1900\,ft^2 * \frac{7.48\,gal}{ft^3} * 0.8 = \mathbf{2,500} \frac{gal}{March}$$

Therefore, this 1,900 square foot house in Columbia, Missouri can collect a potential 2,500 gallons in the month of March.

SI Units Example

Calculate the collection potential (in liters) for a 100-square-meter house with concrete roof in Santo Domingo, Dominican Republic for the month of July.

Looking up efficiency of the collection surface for a concrete roof yields an 0.9, and average rainfall for July is 145 mm:

$$V = R * A * k * e$$

$$V_{July} = 145 \frac{mm}{July} * \frac{1\,m}{1000\,mm} * 100\,m^{\underline{2}} * 1000 \frac{liters}{m^3} * 0.9 = \mathbf{13,050} \frac{liters}{July}$$

Therefore this 100-square-meter house in Santo Domingo, Dominican Republic can collect a potential 13,050 liters (3,447 gallons) in the month of July.

Rules of thumb

The rule of thumb of 0.5 gallons of rainwater caught per inch of rain on each square foot of catchment is quite useful when doing quick calculations in the field.

Equation 13
Rainwater collection potential rule of thumb in Imperial Units.

$$Volume(gal) = 0.5 \frac{gal}{inches * ft^2} * rainfall(inches) * Area(ft^2)$$

The 0.5 gallon per square foot per inch of rain in the rule of thumb can be shown to assume about 0.8 roof efficiency, with the following equation:

$$1in * \frac{1\,ft}{12\,in} * 1\,ft^2 * 7.48 \frac{gal}{ft^3} \cdot 0.8 = 0.499\,gal$$

The calculation in SI units is much easier and more precise. It is the roof efficiency (e.g. 0.8) in liters for every mm of rain on every m² of roof.

Equation 14
Rainwater collection potential rule of thumb in SI Units.

$$Volume(liters) = roof\ efficiency * rainfall(mm) * Area(m^2)$$

This calculation is simple due to the convenience and synchronization of SI units, as shown below:

$$V = R * A * k * e$$

$$Volume = 1\ mm * \frac{1\ m}{1000\ mm} * 1\ m^2 * 1000\frac{l}{m^3} * 0.8 = 0.8\ liters$$

Notice how the *m* to *mm* conversion cancels with the *liters* to *m³* conversion.

4.6 Usage Versus Catchment and Storage

Many locations can collect sufficient water to meet needs. Often the limiting factor ends up being storage volume. Calculating monthly collection capacities minus monthly demand shows how much storage is needed. The spreadsheet in Figure 4-4 shows the monthly rainfall, collection, usage, and storage calculated for a house in Columbia, Missouri with a footprint of 541 ft², a roof efficiency of 0.8, and a storage tank of 800 gallons. Note that this system, assuming it starts full in January the first year, is sustainable with the assumed monthly usages. The tank always has some water remaining, with February of the second year being the lowest at 98 gallons.

Sizing a Rainwater Tank based on Climatological and Usage Data

	User Input	Units	SI Units
Footprint	541	ft²	50 m²
Roof Efficiency	0.8	material dependent	
Size of Tank	800	gal	3028 liters

ID # 231773	Jan	Feb	Mar	Apr	May	Jun	Jul	Aug	Sep	Oct	Nov	Dec
Rainfall (in)	1.21	1.3	2.64	3.62	4.84	4.19	4.17	3.59	4.23	3.4	2.97	1.93
Collection Capacity (gal)	326	351	712	977	1306	1130	1125	969	1141	917	801	521
Usage (gal)	500	500	400	500	700	600	1000	800	400	600	700	900
Tank Vol. starting full in Jan (gal)	800	651	800	800	800	800	800	800	800	800	800	421
2nd year tank volume (gal)	247	98	410	800	800	800	800	800	800	800	800	421

Figure 4-4

Monthly rainfall, collection, usage, and storage calculated for a house in Columbia, Missouri with a footprint of 541 ft², a roof efficiency of 0.8, and a storage tank of 800 gal. This spreadsheet is downloadable at appropedia.org/Basic_rainwater_collection_calculations.

When storage is the limiting factor, it is often due to the available size, budget, or weight durability of the storage area. Dialing back usage and increasing collection areas are two ways to lower the needs of the storage.

5. Full Systems and Stories

The technology is often the easiest part of any project. How people work together can often pose more of a challenge. Rainwater catchment systems provide countless opportunities as well as technical and social challenges. In all of these systems, the most important thing we build is trust.

Built rainwater harvesting systems can provide water for gardens, toilet flushing, hand washing, household use, community use, landscape irrigation, commercial building use, or—with proper purification—even drinking. The following projects are examples of custom, dry, gravity-fed, built systems for single garden irrigation to a small community scale. They are also, more importantly, examples of how communities came together to meet their needs with their resources:

◆ Community Scale in Chiapas, Mexico

◆ Water for School Use in La Yuca, Santo Domingo, Dominican Republic

◆ Daycare Garden in Parras de la Fuente, Mexico

◆ Water for a School Garden in Eureka, California, USA

◆ Democracy Unlimited, Eureka, California

◆ More Systems

Storage Volume
19,000 liters (5,000 gal)

Population served
Small community

Roof Type
Corrugated metal

Usage
Drinking, washing

Catchment Area
150 m² (1600 ft²)

Date first created
2010

Yearly potential
129,000 liters (34,100 gal)

LOCATION

Chiapas
Mexico

5.1 Community Scale in Chiapas, Mexico

A 20-hour bus ride from the Yucatan Peninsula in Mexico brought me into Chiapas, the poorest state in all of Mexico. The capital of Chiapas is Tuxtla Gutiérrez but, in many ways, the cultural capital is San Cristóbal de las Casas. The meandering cobblestone streets are replete with coffee houses, restaurants, custom movie theaters that fit anywhere from two to 40 people, and dozens of languages from all around southern Mexico and the world. In the streets of San Cristóbal, it is hard to remember that extreme poverty is a daily reality for many of its inhabitants. The young indigenous children selling beautiful and sometimes magical clay and felt *animalitos* are shy and well nourished. The loud party music rocking the center of town at night attracts a reasonably sized crowd, and the street vendors seem to sell well. Yet, like many cities, just scratching the superficial layer reveals extreme poverty in its crevices and edges. Interestingly, in the valley of San Cristóbal, much of the poverty is up the mountain, yielding a juxtaposition of roughhewn wood and mud homes with no electricity, no running water, and phenomenal views.

I had come hoping to work with *Las Abejas*, a pacifist contemporary of the *Zapatistas*, and to that end I was to meet with the local non-governmental organization (NGO), Otros Mundos (OM). We were meeting to determine if we were going to work together and, if so, toward what goals and in what capacities. I was excited and nervous. I had just spent the last five years working in Northern Mexico, in the desert of Coahuila. Coahuila is as different from Chiapas as two states could be. In my opinion, if not for political lines, these two states would be in different Americas, with Coahuila part of North America (as all of Mexico is), and Chiapas—with its jungles, animals, foods, indigenous cultures, and mountains—part of Central America like its neighbor Guatemala. In Coahuila, we had already built friendships, partnerships, projects, and trust. Here in Chiapas, it felt like I was starting at step zero, and, with my terrible Spanish, that can be a scary and exciting place to be.

I did have one connection to Otros Mundos through a colleague who had inspired me with their work.[30] Otros Mundos was run by three amazing, change-making women: Tania González, Ursula Lascurain, and Claudia Ramos. Together they possessed a potent mix of backgrounds, including local, indigenous, and educated in DF (Mexico City). My colleague had described their presence and perspicacity. I was not disappointed. The meeting between Otros Mundos and me started in my favorite manner... over food. After getting to know each other, we dove into details of each other's work.

The OM staff described their experiences teaching solar and water technologies in communities throughout Chiapas. They shared their work co-creating an inspiring collaborative agreement between all local NGOs to be safe contacts for reporting domestic abuse. They also shared their involvement in protests against the Mexican government and corporations that were trying to forcibly take indigenous land that had been found to have minerals beneath it. They described their involvement in protests against these same groups' destruction of our future through excessive greenhouse gas emissions.

I described our work, an interesting challenge made more difficult by the fact that even after five years we still had no name.[31] I chronicled our past projects, successes, and failures. I could feel Otros Mundos relaxing while I described with candor the ways in which we had failed before. I made a point to speak to the ways in which we could work together, and how we would love to collaborate on clean water, energy, and food projects, but wouldn't be able to collaborate on protests. I clumsily tried to share how I did not feel it was appropriate for new foreigners to engage in political protests, that in Mexico it was illegal for visitors to engage in protests,[32] and how it wasn't really what our group did.

---

30 That colleague was Alex Eaton. He had helped found a school with Otros Mundos, and he went on to found the very effective Sistema Biobolsa that builds appropriate biogas digester systems.

31 When reporters would ask, I would say we didn't have a name, we just had our work. . . those reporters would usually look at me like I was quite foolish.

32 "Foreigners may not in any way participate in the political affairs of the country." Constitution of Mexico

Specifically, I tried to describe how vital I feel it is that there are people out there telling people what is wrong and what needs to stop; however, our work was in something I called "point positive design." [33] In point positive design, we focus on building better alternatives so that people have healthier options. For example, instead of protesting a new coal factory, we build solar powered systems. Otros Mundos suffered my Spanish, and my discomfort, as I tried to honor their work and describe ours. Eventually, Tania González stopped me, and said, *"Oye, nosotros entendemos, ustedes no son activistas, son practivistas."* That is, "We got it, you are not activists, you are practivists."... and now, after all these years, we finally had a name. Practivistas.

After our day together, we decided to prototype our working relationships. To do that, the following week we collaboratively ran a small needs and resources un-conference. Attendees included local community members, rural community members, community leaders, NGO leaders, and Otros Mundos staff and interns. I presented for 20 minutes on past Practivistas (our new name) projects, and then we broke off into separate tables to discuss the most pressing needs and available resources in the area. We then came back together to brainstorm possible ways to meet current needs with available resources. It was a great time, and deeper relationships were forged along with some great ideas for projects. These projects included a demonstration side-by-side comparison of improved cookstoves so rural community members could comparison-shop the style of system they want; a biogas digester at a local appropriate technology demonstration home with six pigs and a small restaurant; and rainwater harvesting in a rural Tzotzil community.

After reflecting on our un-conference process and results, the next step was to meet more members of a local Tzotzil community, populated by a few families and located a few hours outside of San Cristóbal de las Casas. There we would meet to assess the interest in working together. The Tzotzil are an indigenous Maya people of the state of Chiapas, and the Tzotzil language (*Bats'i k'op*) is a Maya language endemic to Chiapas. I had been instructed that the Tzotzil were generally more reserved with outsiders than many other Mexican communities I had worked with. [34]

---

33 I stole the concept of point positive design from kayaking culture, where you point not to the obstacle but instead to the safer, more desirable, path. Instead of telling people what not to do, make it easy for them to do the desired action.

34 Except for the Rarámuri in the northwest.

The community had already been working with Otros Mundos, so the trust was already there. The reception was warm and so was the food and tea. The matriarch of the house we met in served us eggs and fresh pole beans scrambled together. The ingredients were all from within 100 feet of her house. Her house was home to four children and four adults. It had a red clay floor, roughhewn wood plank walls, and a somewhat effective improved cookstove. After the food, she presented us with lemongrass tea, also from within 100 feet of her house. The tea was incredibly delicious. It was also a safe way to hydrate, as waterborne illness is a serious risk, especially in Chiapas, which has a mortality rate from diarrhea among children under five years old that is three times higher than the Mexican average.[35]

Between watching my surroundings, listening to stories, and entertaining two of the most curious children, it took me a while to understand why the tea tasted so good. The taste was imparted from the smoke of the fire, rendering it smoked lemongrass tea. Looking up, I saw the roof looked almost painted black with soot. Unfortunately, this delightful taste betrayed a nasty truth: indoor air pollution from cooking on stoves indoors is one of the largest killers of women and children in Southern Mexico.[36]

After we ate, drank, and got to know each other, the community leaders invited me to a community meeting the next day. This is a rebel community, and accordingly there are few photos or participant names we can share (and definitely no GPS coordinates). The next day, we made the few-hours journey from the city back to the community. Before we went to the meeting, we stopped in at the matriarch's house, where she fed us again. It always amazes me how often the people with the least, share the most.

At the meeting, I experienced a great "first" in a community meeting. The community leaders, Otros Mundos, and I were gathered in a round room with a mud floor and no stove. The head community leader was speaking about the importance of community

35 Gutierrez-Jimenez, J. (2014). Evaluation of A Point-Of-Use Water Purification System (Llaveoz) in a Rural Setting of Chiapas, Mexico. Journal of Microbiology & Experimentation, 1(3). doi:10.15406/jmen.2014.01.00015

36 Maldonado, Iván Nelinho Pérez, Lucia Guadalupe Pruneda Álvarez, Fernando Díaz-Barriga, Lilia Elisa Batres Esquivel, Francisco Javier Pérez Vázquez and Rebeca Isabel Martínez Salinas (2011). Indoor Air Pollution in Mexico, The Impact of Air Pollution on Health, Economy, Environment and Agricultural Sources, Dr. Mohamed Khallaf (Ed.), InTech, DOI: 10.5772/19864

engagement and trust, and about some serious issues of sanitation, respiratory infection, water, and energy. Partway through his talk, he fell asleep.

There I was, sitting and listening (arguably the most important engineering design skill), and he just nodded off mid-sentence. I worried about his health and felt my own awkwardness. As I glanced around the room, I saw that no one else seemed alarmed. We sat together in silence for a few minutes, until it became natural, and eventually the main community leader woke up and continued speaking.

This small village contains a wealth of nature and community; however, it is resource restricted in money, energy, and water. During the long dry season, the local people must collect surface water for household uses, which results in waterborne illness. Indoor air quality (from the cookstoves) and waterborne illness became our top priorities here. Lifesaving technologies are often mundane technologies. Improved cookstoves can mitigate the health impacts of cooking with biomass by being more efficient and directing the smoke outdoors via a chimney. Rainwater harvesting systems assist in avoiding waterborne illness by catching the water before it touches the ground.

That summer, Practivistas came back with twenty students to work with Otros Mundos, a demonstration home in San Cristóbal, and the Tzotzil community. In the community, we worked together to build improved cookstoves and to design and build a unique rainwater harvesting system.

The main criteria of safety, cost, locally repairable, and water availability through the dry season for the entire community provided some very interesting challenges. Those challenges were made more interesting by the steep and tight topography of the area, the lack of financial resources, and the communications barriers presented by having Spanish as the second language for both the students, who spoke English, and the community members, who speak Tzotzil. Luckily Otros Mundos played a critical role of trust, context, and training.

The design and construction process was a joy. While measuring the roof area, my tape measure broke and the seven-year-old granddaughter of the matriarch helped me fix it. The skills and trust of all the participants together created lasting systems and creative solutions. For instance, in order to utilize more locally available and low-cost resources, we made gutters by bending corrugated roofing metal into a trough. My favorite innovation was to collect and combine the rain from two different houses (Figure 5-1) into one 19,000-liter tank for storage and community use (Figure 5-2). This combination allows for more catchment area as well as for a large centralized storage, at a lower cost than separate tanks. It is a solution that is born from the human-centered design process and the communal nature of the participants.

Figure 5-1

Rainwater harvesting on one of two houses that feed a 19,000-liter ferrocement tank that supplies water for this small community in Chiapas. appropedia.org/Practivistas_Chiapas_rainwater_catchment.

Figure 5-2
19,000-liter ferrocement tank before completion, showing a piece of bent corrugated roof metal used as a gutter in the background. appropedia.org/Practivistas_Chiapas_rainwater_catchment

Storage Volume
2,000 liters + cistern

Population served
School

Roof Type
Corrugated metal

Usage
Bathrooms, cleaning

Catchment Area
16 m² (172 ft²)

Date first created
2011

Yearly potential
22,000 liters (5,800 gal)

LOCATION

Santo Domingo,
Dominican Republic

5.2 Water for School Use in La Yuca, Santo Domingo, Dominican Republic

In 2010, Practivistas Chiapas (the summer abroad program) was mandated by the California State University system to move out of Mexico because of the drug war. In search of new locations, I landed in Santo Domingo, the capital of the Dominican Republic. The Dominican Republic shares the island of Hispaniola with Haiti. My first night there was filled with what have become sounds of comfort for me: the streams of music, especially *bachata* (a Dominican bitter-yet-romantic couples dance music) from every *colmado* (very social corner stores), and the clack of dominoes.

Walking around the city, I met many interesting locals and one angry Spaniard. He told me how Dominicans didn't care about the environment. I told him that I was going to an event being held by the local 350.org group the next morning. The event was to raise awareness for climate change, especially from the viewpoint of an island country that will be heavily affected by sea level rise. The event was a great success and DR was the only island country to be satellite photographed for the global 350.org event (Figure 5-3). At the event, I got to know *Colectivo Revark*, a non-profit founded by the intrepid local architects Wilfredo Mena Veras, Abel Castillo Reynoso, and Joel Mercedes Sanchez. We eventually became close friends and colleagues, but first we built a relationship based on trust, work, and dancing. Colectivo Revark had just held an "Architecture for Earthquake Disaster Relief" competition, called *Sismos 2.0*. We worked together to judge the submittals. After that successful experience, we built the relationship from afar (I returned to the US), and started working on developing the Practivistas Dominicana Program.

Figure 5-3
Satellite image in 2010 for 350.org of Dominicans (plus Lonny Grafman and Gabriel Krause in the top left hand) holding umbrellas and arranged in the form of a house being consumed by rising sea levels (left credit: DigitalGlobe). Participants raising awareness about the threats of sea level rise just one week before the COP 16 climate change conference (right credit: Marvin del Cid). flickr.com/photos/350org/ CC BY-NC-SA 2.0

The next year, in 2011, we returned to work with Colectivo Revark and study at Universidad Iberoamericana (UNIBE) under the care of the Director of Architecture, Elmer Gonzalez. Outside of the university, Colectivo Revark was our community liaison and was critical in all the community engagements—yet our community engagement in La Yuca almost did not happen.

La Yuca is one of the most financially poor urban barrios in the center of Santo Domingo, Dominican Republic. La Yuca contains one school, whose schoolyard also serves as the hub of activities for children. You can touch the walls of houses on both sides of the narrow streets as you make your way through the serpentine labyrinth of the neighborhood. If a moped is coming through, you need to press yourself out of the way. The neighborhood bustles with constant activity and noise, including the comforting endemic rhythms of *bachata* and dominoes. Overhead, masses of makeshift grid electricity extend into the crevices of every home. These floating balls of electrical spaghetti are worked on by many people of La Yuca as repurposed wires burn out in spectacular light shows and need to be replaced.

The story we first heard from a beautiful ninety-year-old *abuela*[37] is that La Yuca was originally a temporary home settled by the workers who built much of the surrounding parts of the city, and then the workers didn't leave. There have been many attempts to push La Yuca and its inhabitants out of the city, but La Yuca has prevailed.

Colectivo Revark and UNIBE helped set up our first community meeting with the *junta de vecinos* (city council) of La Yuca. During that first community meeting, the reception was low energy and the junta de vecinos seemed mostly uninterested in working together. It was only after the pastor understood what we were proposing and restated it with eloquence, that the engagement happened. He reiterated that we were not there as a charity organization. We were not there with a "solution." We were not there to drop something off and take pictures. We were there to work and learn together. We were there to seek solutions together. And we were there so that we could all gain knowledge, capacity, and build a better future together.

After that initial meeting, we decided to have a meeting open to the entire community where we could identify our top needs and resources. Some top needs included clean water (some people were spending over 40% of their income on water), more school space (there are more students than can fit in the school), electricity (11% of Santo Domingo re-appropriates their electricity), and jobs (incomes are often just a couple of dollars a day).

The open community meeting was loud and fruitful, especially due to the deep support of the town mayor, Osvaldo de Aza Carpio.[38] We brainstormed dozens of available resources and top needs. Then we prioritized the top needs into just a few and broke off into small groups to brainstorm solutions to those top needs... but suddenly, the previously

---

37   Kiva Mahoney (the assistant director of Practivistas Domincana) searched the neighborhood to learn its history and first met this amazing abuela (grandmother/elder).

38   Osvaldo was not only the consummate community leader in La Yuca - in the years following this first engagement he has been instrumental in meetings in other communities and has continued to be a sustainability practitioner and advocate of building technologies in his community and the communities of others.

loud meeting became quiet. With the help of our community partner Colectivo Revark, I realized that an assumption I had made was about to weaken the community meeting. When working in a rural environment, I keep in mind that many of the participants may not be comfortable writing in front of others. I missed that that might be true as well in the urban environment of Santo Domingo. Quickly we changed the break-off groups so that one person, someone we knew would be comfortable writing in front of people, was writing in each group. The noise and cacophony of creation came back up and the results led to years of engagement.

Together we decided to create a wind and solar renewable energy system, a schoolroom from plastic bottles (a style called ecoladrillo), and a rainwater harvesting system on top of the new schoolroom we were building together. The school was currently ordering two trucks worth of water per month, which was expensive and was only enough water to clean the school and to flush the toilet manually at night. The school had no water for hand washing in the toilet nor anywhere else (a major health indicator).[39]

Together we built a 2000-liter rainwater harvesting system that also included additional storage in an existing cistern; however, the rainwater project at the grade school in La Yuca took a few years of iteration to get right.

The first year, it caught water . . . but the water wasn't really trusted by the users because the slow-sand filter looked too "rural" to the users. Therefore, we replaced the slow-sand filter with more urban-looking canister filters.

Another problem was that the original gutter system, which consisted of PVC sliced open and pressed over the edge of the corrugated roof, was too flat and was getting clogged, causing it to sag and leak. In 2014, we replaced the PVC with a new, more conventional K-type gutter, which fixed the sagging and leaking problem (Figure 5-4). We also added a fabricated splash guard to protect the very close neighbors from splashing during heavy rains.

---

39 Handwashing: Clean Hands Save Lives. (2015, November 18). Retrieved from https://www.cdc.gov/handwashing/why-handwashing.html

Figure 5-4
Sliced-open PVC gutter (left) replaced in 2014 with a conventional gutter and fabricated splash guard (right). appropedia.org/La_Yuca_rainwater_2014

Other problems included the speed of water delivery, the faucets being too tall for the director, and the bathroom not being integrated. All those problems have been resolved through various adaptations described in the Components section.

The rainwater harvesting system (Figure 5-5) installed at the grade school provides hand washing for the students (Figure 5-6) for the first time in over a decade, plus a shower for the director and water for cleaning the school. Since the installation of the rainwater harvesting system, the school reduced their order of two trucks of water per month to just two trucks per year, for a savings of 22 trucks of water per year—year after year.

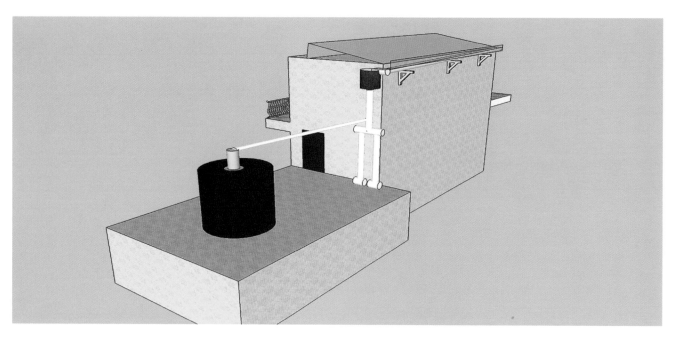

Figure 5-5
Updated rainwater harvesting system provides the water for cleaning, handwashing, toilet flushing, and showering in La Yuca, an urban barrio of Santo Domingo, Dominican Republic. appropedia.org/La_Yuca_rainwater_2014

Figure 5-6
Rainwater providing a shower for the director (upper left with Oswaldo and Tanja), storage in the cistern (lower left), and handwashing (right) in La Yuca an urban barrio of Santo Domingo, Dominican Republic. appropedia.org/La_Yuca_rainwater_2014

Storage Volume
400 liters (106 gal)

Population served
Daycare families

Roof Type
Concrete

Usage
Small garden

Catchment Area
22.9 m² (246 ft²)

Date first created
2008

Yearly potential
6,045 liters (1,600 gal)

LOCATION

Coahuila
Mexico

5.3 Daycare Garden in Parras de la Fuente, Mexico

Parras de la Fuente is an oasis in the desert of Coahuila in Northern Mexico. As its name (grapevines of the fountain) suggests, Parras is replete with ample spring water and the first winery in the Americas. That said, the community members we worked with in Parras were very concerned about the threat of water privatization and water pollution from local industry. This political and environmental context was explored through my co-director, Dr. Francisco de la Cabada,[40] and our community liaisons from Universidad Tecnológica de Coahuila (UTC), especially through the work of Carlos Alejandro Ramírez Rincón (director) and Simón Leija (environmental science instructor).

In the summer of 2008, students in the Practivistas program worked with a local daycare and center for women and children to build a small, 9-square-meter school garden and a rainwater harvesting system (Figure 5-7) to water the garden.

As the daycare facility focuses on education and is quite public, a criterion for success— in addition to effectiveness, safety, and aesthetics—was educational value. To meet that criterion, the system was labeled to inspire and teach the visitors about water conservation and collection.

The summer months are the most active for the garden, yet incur the least rainfall, so sizing the system to have enough storage in July was a major constraint. One important lesson learned in this system was that the first donated storage tank had been reused to store an unknown chemical. While the plastic was originally food grade, we were not able to identify the specific contaminant and therefore the tank was rendered unusable. A new, food grade tank was secured before we completed the system.

---

40 Dr Francisco de la Cabada teaches Spanish at Humboldt State University and is from Coahuila. His connections to the land, the art, and the people founded a rich engagement in Parras.

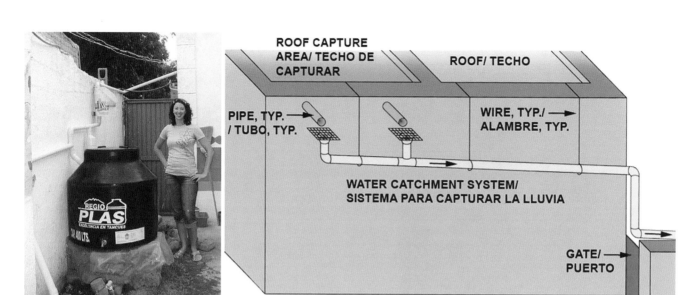

Figure 5-7

The water tank, piping, filter, and first-flush (left), and the capture area, filters, gutters, and piping (right) at a daycare in Parras de la Fuente, Mexico. appropedia.org/Daycare_rainwater_catchment_system

Storage Volume
1,892 liters (500 gal)

Population served
School

Roof Type
Tar and gravel

Usage
Garden, grounds

Catchment Area
83.6 m² (900 ft²)

Date first created
2014

Yearly potential
65,900 liters (17,400 gal)

LOCATION

Eureka
California, USA

5.4 Water for a School Garden in Eureka, California, USA

For ten years, I tried to partner my Engineering 215: Introduction to Design sophomore engineering course with a public grade school. As someone who was mostly failed by the U.S. public school system and who dropped out before my sophomore year, I am acutely interested in public education. But public schools are typically risk-adverse, and year after year I heard that such a partnership was unlikely.

Instead, we partnered with non-profits that were willing to take a risk on sophomore level university students. We completed hundreds of projects with these partners, and eventually landed some charter schools (which can be nimbler than traditional U.S. public schools) to start building trust in collaboratively built education.

Finally, in 2014, a brave counseling services director, Trevor Hammons, and principal, Jan Schmidt, from Zane Middle School decided to take a chance on us. Zane Middle School is in Eureka, California and serves youth from grades six through eight. More than three-fourths of its students are on an assisted lunch program. Zane looked just like the schools from my childhood. Concrete, steel, asphalt, fence and dirt engendered the feel of a prison. The outside had that cold industrial feel, as if students are supposed to be uniform products of an unyielding machine, but inside the cold stark walls, inspired teachers and future-thinking administrators bustled.

That year, we took on nine projects to build learning apparatuses and to transform their concrete and metal campus into a more natural, inviting, and thriving campus. Students worked directly with teachers to build projects such as a songbird refuge, an upgraded garden, edible landscaping, and a rainwater harvesting system.

The rainwater harvesting system was needed to support the gardens and edible landscaping, educate students on sustainable water practices, and be another source of emergency water for natural disaster preparedness (Eureka is in a seismically active region). Together the students worked with staff and teachers to determine the top criteria of safety, ease, durability, cost, and educational value. The students designed a robust 500-gallon dry system with easy access to the gardens and edible landscaping (Figure 5-8 and Figure 5-9). However, after reviewing prototypes, the teachers and staff found a critical problem, a problem that would either break the system in less than a year or cause other issues.

The problem was with the conveyance system, which used a pipe leading from the gutter to the tank. The pipe would be high, but low enough that an enterprising youth could jump up and grab onto it in a covert attempt to access the roof. The team members and staff reviewed various ways to mitigate the problem, such as making the pipe less appealing, creating a wet (instead of dry) system where the conveyance pipe would be buried below ground, and moving the storage against the wall. All these solutions were less desirable with regard to durability or ease. Finally, the team found an exciting and simple innovation.

The innovation was a zip-tie and a flex fitting. The zip-tie connects the conveyance pipe to the gutter and a flex fitting at the other end of the pipe allows for bending (Figure 2-9). So now, if a student jumps up and grabs the pipe, instead of the gutter or roof breaking, a 5-cent zip-tie breaks. Over the last four years, the zip-tie has needed to be replaced only three times. This type of important but simple innovation only comes from working closely with your client and constituents, building empathy, and understanding the users.

That year, the local media quoted the principal as saying:

> *"I am very impressed with the quality of the projects designed by these college sophomores. We look forward to deepening our partnership with HSU in the coming years."*
> *~Jan Schmidt, Principal*

That is exactly what has happened. These projects have been an incredible success and have led to years of HSU-Zane partnerships and a transforming campus.

Due to Zane Middle School's push toward a more inspirational STEAM (Science, Technology, Engineering, Arts, and Math) school, including Zane's dedication to working with HSU students to build sustainable infrastructure, Zane received the Gold Ribbon Award from the California State Board of Education.

Figure 5-8

Rainwater harvesting system at Zane Middle School in Eureka, California used for the school garden. This photo was taken three years after installation. appropedia.org/Zane_Middle_School_rainwater_catchment

Figure 5-9
Zane Middle School garden with the rainwater harvesting system in the background (left) and during a presentation (right). appropedia.org/Zane_Middle_School_rainwater_catchment

Storage Volume
9,464 liters (2,500 gal)

Population served
Community org.

Roof Type
Asphalt shingles

Usage
Garden, chickens

Catchment Area
50.3 m² (541 ft²)

Date first created
2007

Yearly potential
42,100 liters (11,100 gal)

LOCATION

Eureka
California, USA

5.5 Democracy Unlimited, Eureka, California

Democracy Unlimited of Humboldt County (DUHC) was a local community organizing space in Eureka, California. DUHC hosted monthly pancake breakfasts and helped to launch both the Humboldt Independent Business Alliance and the Humboldt Community Currency project. It also launched Move To Amend, a national campaign to abolish corporate personhood.

I first met the director of DUHC, David Cobb, at a community meeting I had organized in 2006 to look for community-based solutions for the problems facing our Northern California communities. I was impressed by his clarity and commitment to community action, and was not surprised to learn that he had been the 2004 Green Party presidential candidate. In 2007, I was excited to learn that DUHC wanted a rainwater harvesting system to model resilient community technologies and provide water for their gardens, chickens, and landscaping.

That year, DUHC became the client for my Engineering 305: Appropriate Technology course, where student Nicole Vincent took on the task of collaboratively designing and building a rainwater harvesting system for the DUHC house. The design was fairly straightforward, as the house had two stories—providing ample head—and already had gutters. The roof area was sufficient to collect enough water to last through a three-month dry season by storing the water in a 2,500-gallon tank.

The main design constraint became the topography of the site; it was quite flat. Careful analysis showed that there were only three options to get the rainwater from the storage to the gardens and chicken coop. The three options were: to raise the tank so that water could flow by gravity from the roof to the tank and still have enough pressure to reach the end use; to put the tank on the ground in the back yard and use a pump to pressurize the water enough to reach the end use; or to put the tank in the slightly higher front yard.

Raising the tank provided a significant financial and engineering hurdle, as 2,500 gallons is 20,900 pounds . . . over 10 tons! Additionally, using grid energy to pump the water did not meet the goals of DUHC, and solar power in 2007 was still quite expensive. For those reasons, the front yard became the home for the tank (Figure 5-10).

Figure 5-10
The DUHC house with 2,500-gallon rainwater tank (left) with an end use of chickens and duck (center) and gardens (right).

Nicole and volunteer helpers constructed the system. With the first rain, the water flowed, the tank started filling, and the piping had enough pressure from gravity for the end use. Unfortunately, we made one large mistake.

Without realizing it, the rainwater tank was in violation of the Eureka set-back laws. If we had realized that ahead of time, we could have petitioned—but instead, the city wanted us to take it down. That started a one-year push to change the law of Eureka. We had luck on our side. David Cobb was a trained lawyer, and the City of Eureka general plan language was heavily weighted toward supporting this type of gravity-fed, natural, water-saving technology (Figure 5-11).

Wherever feasible, the natural terrain, drainage and vegetation of the neighborhood should be preserved with superior examples contained within parks or greenbelts.

Neighborhood design should help conserve resources and minimize waste.

Neighborhoods should provide for the efficient use of water through the use of natural drainage, drought tolerant landscaping, and recycling.

Figure 5-11
City of Eureka - General Plan Goals & Policies in 2007 – Policy 1-K-1.

Members of the community center came together to prove that the tank was located in the best space to conserve resources, and that a system like this deserved a variance. A local engineer-in-training, Tressie Word, took the lead on site mapping, enlisting the help of students and community members, and a DUHC leader, Kaitlin Sopoci-Belknap, took the lead on language and education. In 2008, through the labor of this collaboration, a variance was granted. This variance means that anyone in Eureka can install a rainwater tank wherever it makes the most sense for gravity, regardless of the setback. The next year, Kaitlin and others started teaching rainwater harvesting classes and the local media helped promote the idea (Figure 5-12).

Saturday, Jan. 17, 2009 — Times-Standard — B5

GREEN LIVING

Saving water in a big way

Sharon Letts
The Times-Standard

Eureka residents Kaitlin Sopoci-Belknap and David Cobb are no strangers to human ecology.

Sopoci-Belknap earned a bachelor of arts degree in culture, ecology and sustainable community from the New College of California. Her partner, David Cobb was the Green Party nominee for President of the United States of America in 2004. Both are principals with Democracy Unlimited, a non-profit that designs and implements grassroots strategies exercising democratic power over corporations and governments.

So, it made sense when the couple decided to save a little rain water for their garden and fowl at their Eureka home.

Everyone is served with a rain catchment system, and in this house chickens and a vegetable garden will be equally well-watered. Shown here are homeowner's David Cobb (middle), Kaitlin Sopoci-Belknap (right) and housemate, Megan Wade Antieau (left), with chickens.

Figure 5-12
The local Eureka, CA news, Times-Standard, covering the rainwater story and promoting upcoming workshops in 2009.

If it wasn't for the capacity of the local community—including law, engineering, education, and more—this system would have been decommissioned. Now, ten years later, it has inspired (and continues to inspire) other systems and water conservation education in general. As DUHC's director said:

"Harvesting rainwater has saved us money, reduced our environmental impact, and helped us to take a small step towards being more sustainable and resilient."

~David Cobb

5.6 More Systems

Dozens of projects are documented on Appropedia. These projects usually cover background, literature review, criteria, constraints, design, construction, maintenance, and testing. In addition, some of them have follow-ups from years later discussing successes, failures, and evolution.

These innovations and implementation are all shared, open source, to build upon, adapt, and improve rainwater harvesting for all at http://appropedia.org/Portal:Rainwater_harvesting. If you are looking for DIY examples to follow, this is the place to look (Figure 5-13). Some examples include:

- Rainwater for wine barrel soaking
- Integrated, permitted, rainwater for household-wide use in Portland, OR, USA
- Rainwater for a hostel in Costa Rica

Example Rainwater Harvesting Systems from Appropedia

Figure 5-13

Many more example systems, replete with research, criteria, constraints, budgets, timelines, failures and successes, at appropedia.org/Portal:Rainwater_harvesting.

5.7 Organizations

I love to hear the amazing stories of communities and individuals coming together to meet their needs with their resources. I also love to meet organizations that help make that happen. The following two organizations, Isla Urbana and Paul Polak's Spring Health, exemplify widely deployed, small-scale rainwater harvesting for a better future.

5.7.1 Isla Urbana

Isla Urbana designs and builds their systems in Mexico City and is one of the many amazing organizations working on rainwater harvesting systems around the world. They work with many partners to teach, train, build, and install rainwater harvesting in Mexico City and communities throughout Mexico.

As of this writing, Isla Urbana has installed more than 5,000 systems that have together captured over 360 million liters of rainwater. They have installed 400 of these systems in rural communities and the remaining 4,600 systems in the city. (Figure 5-14 and Figure 5-15)

Figure 5-14
Rainwater harvesting system at a home in southern Mexico City with a Tlaloque first-flush by Isla Urbana.
Provides six months of total water autonomy per year. appropedia.org/Isla_Urbana

Figure 5-15
Rainwater for dishwashing in Mexico by Isla Urbana. appropedia.org/Rainwater_catchment_at_Isla_Urbana

Through an iterative design process, Isla Urbana has designed a simple and sophisticated system that accounts for usage patterns, leaf litter and debris, and first-flush through a device they call the Tlaloque. Their systems also include a "calm inlet" (Figure 5-16 point

3), which slows down water entering the storage tank to avoid disturbing any sediment that may have collected at the bottom of the storage tank. Finally, Isla Urbana's systems include automatic chlorination and filtration (Figure 5-16 Point 4 and 6) to keep water safe and low maintenance.

Figure 5-16
System and components for an Isla Urbana rainwater harvesting system. islaurbana.org

Another aspect of Isla Urbana's approach that is worth learning from is that they focus on neighborhoods. By installing systems close to each other, and filling a neighborhood with rainwater harvesting systems, Isla Urbana builds a critical mass of connected people who know how to operate the systems. These neighborhoods of catchment become a resource to each other, which helps keep systems operating optimally, and they become seed neighborhoods that spread the word to other neighborhoods.

5.7.2 Spring Health

Spring Health sells clean water in India, specifically in areas of poverty. In their mission to "improve the health of millions of poor customers through safe drinking water by lessening sickness, eliminating the costs of expensive medicines and doctor visits, and improving livelihoods," Spring Health takes a unique, market-driven, entrepreneurial approach to rainwater harvesting. Whereas Isla Urbana installs rainwater harvesting systems on homes and organizations, Spring Health employs a base-of-the-pyramid entrepreneurship model to install rainwater harvesting businesses that then deliver and sell the cleaned rainwater at a better-than-market price in communities that make less than two dollars per day. These deliveries are often made using bicycles (Figure 5-17). One of the many surprising results of Spring Health's iterative design process was that financially poor people would pay up to 1/3 more to have their water delivered instead of picking it up. That said, even with the additional delivery cost, Spring Health provides ten liters of clean drinking water, enough for a family of five, for less than $0.10 USD per day.

As of this writing, Spring Health sells drinking water to over 150,000 people in 260 villages every day. Their goal is to reach 100 million customers in India every day, before expanding to other countries such as Bangladesh, Pakistan, and Kenya.

Figure 5-17
A Spring Health rainwater harvesting system (left) and bicycle delivery vehicle (right). http://www.paulpolak.com/_slide/spring-heath/

One of the co-founders of Spring Health is the inspirational Paul Polak. Paul is the author of *Out of Poverty* and *The Business Solution to Poverty*. He has met with over 3,000 financially extremely poor families over the last 30 years, and has built businesses to serve the base-of-the-pyramid. Paul's work is illuminated well in one of his statements:

"90% of the world's designers spend all their time addressing the problems of the richest 10% - before I die, I want to turn that silly ratio on its head."

~Paul Polak

6. Other Useful Stuff

In your quest to more deeply understand rainwater harvesting and competency toward designing and building, you will need to collect many more tools. Following are a few of those tools to help along the way, specifically: *Units Related to Water* and where to find *More Information*.

6.1 Units Related to Water

The following unit conversions are directly related to sizing and analyzing rainwater harvesting systems in terms of length, area, volume, mass, pressure, and general water conversions.[41]

Length

- 1 m = 3.28 ft = 100 cm

- 1 ft = 12 in = 30.48 cm

- 1 mi = 5,280 ft = 1.61 km

---

41 For sources, see, for example: Pressure Units - Online Converter. (n.d.). Retrieved from http://www.engineeringtoolbox.com/pressure-units-converter-d_569.html Rainfall on Roofs and Gutter Slopes. (n.d.). Retrieved from http://www.engineeringtoolbox.com/sloopes-roof-drainage-d_1107.html

Area

- Rectangle: length * width
- Circle: $\pi * r^2$
- $1\ m^2 = 10.765\ ft^2$
- $1\ km^2 = 0.386\ mi^2 = 1,000,000\ m^2$
- 1 hectare $= 10,000\ m^2 = .01\ km^2 = 2.47$ acres
- $1\ mi^2 = 2.6\ km^2 = 640$ acres
- 1 acre $= 4,047\ m^2 = 43,560\ ft^2$ (66 ft * 660 ft) $= 1/640\ mi^2 \approx 40\%$ of a hectare

Volume

- Area * depth
- Rectangular prism: length * width * depth
- Cylinder: $\pi * r^2 *$ depth
- $1\ L = 0.264\ gal = 1000\ cm^3 = 1000\ ml$
- $1\ m^3 = 1000\ L = 35.3\ ft^3 = 264\ gal$
- $1\ ft^3$ (cf) $= 28.32\ L = 7.482\ gal$
- 1 gal $= 3.785\ L = 4$ qt $= 16$ cups $= 128$ fl oz
- 1 acre-foot $= 43,560\ ft^3 = 325,851\ gal$

Mass

- 1 kg $= 1,000$ g $= 2.205$ lb
- 1 lb $= 453.6$ g $= 16$ oz
- 1 US short ton $= 907$ kg $= 2,000$ lb
- 1 metric tonne $= 1,000$ kg $= 2,205$ lb

Pressure

- 1MPa = 10 bar = 9.87 atm = 145 psi
- See water pressure for more

Water (at 4°C and 1 atm)

- Density: 1 g/ml = 1 g/cm^3 = 1 kg/L = 1,000 kg/m^3
- Density: 62.4 lb/ft^3 = 8.34 lb/gal
- Pressure: 0.4335 psi for every vertical foot of water

6.2 More Information

Errata and developments: Rainwater harvesting systems continue to evolve, especially the more we share and study techniques. Also, while every effort has been made to keep this book error free, any errors found will be immediately documented. These next-step evolutions and any errors found will be documented at http://appropedia.org/To_Catch_the_Rain/Errata.

Other books, research, and calculators: Many great books and peer reviewed journal articles cover rainwater harvesting. In order to stay up to date, these books are at http://www.appropedia.org/Rainwater_books. The articles are collected at http://www.appropedia.org/Rainwater_catchment_literature_review. In addition, rainwater calculators are listed at http://www.appropedia.org/Rainwater_calculators.

Climate information: Great places to look for climate data, especially precipitation data, include the United States Geological Survey (USGS), National Oceanic and Atmospheric Administration (NOAA), and Global Precipitation Climatology Centre (GPCC). More up-to-date climate data sets are listed at http://www.appropedia.org/Climate_data.

Project how-tos: The majority of the projects in this book are covered in full detail in Appropedia. Projects are listed and linked at http://appropedia.org/To_Catch_the_Rain#projects.

6.3 Economics

Rainwater catchment systems can vary greatly in price and payback time. Some systems will pay themselves off quickly (e.g., in systems that select inexpensive materials and where clean water is expensive), some systems will take years to pay themselves off (especially in areas where materials are expensive and clean water is very inexpensive), and some systems will pay themselves off immediately where clean water is difficult to obtain. A simple way to examine the costs is to total the construction and maintenance costs and compare them to the replaced water costs (Figure 6-1).

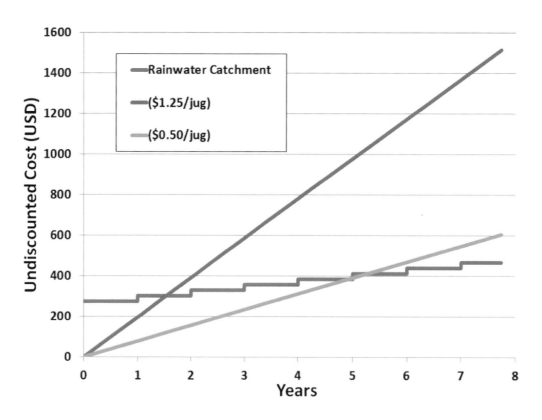

Figure 6-1
Simple payback of the La Yuca system compared to 5 gallon jugs from Jesse Shrader appropedia.org/La_Yuca_rainwater_catchment_2012

In the La Yuca rainwater system, two payback periods were found, one with low-cost water and one with high-cost water. The payback period for the high-cost water (specifically $1.25 per five-gallon jug) was approximately 1.7 years. The payback period for the low-cost water (specifically $0.50 per five-gallon jug) was approximately 5.2 years. The yearly steps up in the rainwater catchment system line represent the filter replacements. These filter replacements are paid back quickly in both the high-cost and the low-cost water scenarios.

Ultimately it is up to the users to determine the value of the system.

6.4 Laws

Rainwater catchment is legal in much of the world. As drought strikes more areas, local laws are changing to often allow, and sometimes encourage or enforce, rainwater catchment. For up to date links and information see http://appropedia.org/Rainwater_laws.

6.5 Disclaimer

All information in this book is provided for informational purposes only. By no means is any information presented herein intended to substitute for the advice which may be provided to you by a professional.

In addition, many locations have specific laws related to rainwater catchment. Please make sure to check your local laws.

The purpose of this book is to inform and to encourage research, collaboration, innovation, and action while building capacity and agency. Please read critically and refer to the many sources provided in the book and on the related Appropedia links. If you use this information at home or anywhere else, exercise extreme caution and utilize the advice of others. Take appropriate precautions for your own safety and the safety of others.

Any email correspondence with author, editors, or publishers of this book is also covered by this disclaimer.

6.6 Cleaning

Different rainwater harvesting systems will necessitate different levels of cleaning. For example, a system made for potable water will require more cleaning than a system made for watering plants. In addition, a dirtier location will need more cleaning than a clean location. Finally, a system with a well-designed screen and first-flush will require less storage cleaning than a system without these measures.

The most common cleaning methods include visually inspecting gutters and screens often for buildup that can be cleaned by hand. Checking every month for the first year, will help you determine what schedule you should be checking at in the future. In addition, the storage should be checked every year or two for buildup. Buildup should be removed by hand, vacuum, or a drain (if one is designed into the system). After cleaning out buildup, check the inside storage surfaces for any scrubbing that may be needed. Finally, depending on your storage tank material, you may consider adding chlorine periodically for cleaning. See appropedia.org/Rainwater_tank_cleaning for more.

7. Problem Sets

The following problems can be used to hone your math skills, develop your sense for the units, practice sizing your own system, for math and science classes, or just for fun. The problems are loosely broken up into Usage, Catchment, Storage, Gutters, First-Flush, Purification, and Synthesis, but you might want to just find the ones that inspire and/or challenge you.

7.1 Usage

1. Convert 10 LPS (liters per second) into a flow rate in GPM (gallons per minute).

2. Calculate the total usage (in gallons) of a sink that flows at 2.00 GPM for 3 minutes.

3. Calculate the flow rate (in GPM) of a sink by using a 5.00-gallon bucket and a timer. In this case, assume that the bucket fills in 150 seconds.

4. Calculate the flow rate (in GPM) of a sink by using a 1-liter bottle and a timer. In this case, assume that the bottle fills in 8.00 seconds.

5. Calculate the total usage (in gallons) of a 1.50 LPF toilet used 4.00 times per day after 1 month (assume March).

6. Calculate how long it will take to fill an 800-gallon tank with a hose that has a flow rate of 10 GPM.

7. Calculate the total volume (in gallons) of water wasted after 1 month (assume March) from a 0.100 GPM leak.

8. Calculate the total monthly (assume March) water usage of a house given the following data:

  ⬥ One 2.00 LPF toilet used twice daily
  ⬥ One 1.50 GPM sink used 15 minutes daily
  ⬥ One 1.80 GPM sink used 20 minutes daily
  ⬥ One 2.30 GPM shower used 8 minutes daily
  ⬥ One 3.00 GPM hose used 10 minutes daily
  ⬥ One 9.00 Gallons Per Wash dishwasher used twice per week
  ⬥ One 20.0 Gallons Per Wash clothes washer used once per week.
  ⬥ Remember to combine each result for the total monthly water usage!

9. Complete a home water usage audit for your own home or, if you do not have a home, consider your daily water usage wherever that is.

10. Capture a video of 2 GPM.

7.2 Catchment

11. Calculate the total volume (in gallons) of 2.00 inches of rain falling into a container with a cross-section of 1.00 ft².

12. Calculate the total volume (in gallons) of 2.00 inches of rain falling into a cylindrical container with a diameter of 1.00 ft.

13. Calculate the total volume (in gallons) of 2.00 inches of rain falling into a container with a cross-section of 800 ft^2.

14. Calculate the total collectable volume (in gallons) of 2.00 inches of rain falling onto a metal roof with a footprint of 1200 ft^2.

15. Calculate the total collectable volume (in gallons) of 2.00 inches of rain falling onto a tile roof with an area of 1500 ft^2 and a footprint of 1300 ft^2.

16. Calculate the total collectable volume (in liters) of 50.0 mm of rain falling onto a metal roof with a footprint of 160 m^2.

17. Calculate and graph the monthly collection volumes (in gallons) for the rainfall in Columbia, Missouri on a metal roof with a footprint of 1200 ft^2.

18. Calculate and graph the monthly collection volumes (in gallons) for the rainfall in your hometown on a metal roof with a footprint of 800 ft^2.

19. Calculate the monthly collection volume (in gallons) of 2.00 inches of rain falling onto 2.00 acres of catchment.

20. Take a photo of 100 square feet.

7.3 Storage

21. List four types of rainwater storage containers.

22. Calculate the storage volume (in gallons) of a 3" PVC pipe that is 1.00 ft long.

23. Calculate the storage volume (in gallons) of a 4" PVC pipe that is 1.00 ft long.

24. Calculate the storage volume (in liters and gallons) of an IBC tote that is 1.00 m x 1.00 m x 1.05 m.

25. Calculate the storage volume (in gallons) of a ferrocement tank that is 8.00 ft in diameter and 10.0 ft tall.

26. Calculate the necessary storage volume to catch all the rain in the highest rainfall month in Columbia, Missouri falling on a tile roof with a footprint of 900 ft^2.

27. Calculate the necessary storage volume to catch all the rain in the highest rainfall month in Columbia, Missouri falling on a metal roof with a footprint of 900 ft^2.

28. Calculate the height needed for a cylindrical ferrocement tank with a diameter of 5 ft to hold 400 gallons of rainwater.

29. Calculate the depth needed for a pond liner lined hole that is 8 ft long by 4 ft wide to hold 500 gallons of rainwater.

30. Take a photo of 100 gallons of storage.

7.4. Gutters

31. List and briefly describe four types of gutters.

32. Use the area rule of thumb method to size gutters for an almost flat 1100 ft^2 roof in Columbia, Missouri.

33. List three ways to correct having gutters that are too small.

34. Calculate the weight, in pounds, of a 3" pipe that is 12 feet long and filled with water.

35. Calculate how much lower the end (outflow) of a gutter should be than the start of the gutter for a 20-foot-long expanse using the minimum rule of thumb.

7.5. First-Flush

36. Describe the purpose of the first-flush.

37. Use the exponential decay method to determine the optimal first-flush volume for a roof with a footprint of 1100 ft^2 in Columbia, Missouri and a λ of 0.8.

38. Use the area-based rule of thumb to determine the optimal first-flush volume for a roof with a footprint of 1100 ft^2 in Columbia, Missouri.

39. Describe an existing method, or create and describe your own method, to capture and evacuate the first-flush water. Draw and label your design.

7.6. Purification

40. Describe the purpose of filtration in a rainwater harvesting system.

41. List and briefly describe the best uses for four different water purification options.

7.7. Synthesis

42. Describe the difference, with pros and cons, of a dry versus a wet rainwater harvesting system.

43. Describe the difference, with pros and cons, of an active versus a passive rainwater harvesting system.

44. List and briefly describe the components of a built household rainwater harvesting system.

45. List and briefly describe the pros and cons of rainwater harvesting from a technical perspective.

46. List and briefly describe the pros and cons of rainwater harvesting from a financial perspective.

47. List and briefly describe the pros and cons of rainwater harvesting from a socio-cultural perspective.

48. What are the top three largest challenges to meeting water needs through rainwater harvesting? Justify your answer.

49. Design your own integrated system for rainwater harvesting. Make sure to consider catchment, piping, first-flush, gravity, storage, purification, and final usage. Draw and label your system.

Afterword

This was my first book. Like many authors, writing this book has been a labor of love for me. I have five trilogies in mind for this same style of book. Each focused on a different resource (e.g. water, energy, and non-physical resources as well). That said, as with all my designs, I want to iterate with you. I also think of this book, as most things in my life, as a prototype. I will write the next book, if this book reaches a certain level of impact. If not, I will probably focus my energy elsewhere. Here are some things you can do to help:

1. Write a review for this book at
   http://appropedia.org/To_Catch_The_Rain#Add_your_review

2. Buy a copy of the book at
   http://tocatchtherain.org

3. Share this book with others

4. Share online with the tag *#tocatchtherain*

5. Share your rainwater story online at
   http://appropedia.org/To_Catch_The_Rain#Your_story

Regardless of any follow-up, please accept my sincere appreciation for your interest, engagement, and efforts to create a better future for your community.

Works Cited

Cobbina, S. J., et al. "Rainwater quality assessment in the Tamale municipality." Int. J. Sci. Technol. Res 2 (2013): 1-10

DeBusk, K., & Hunt, W. (2014, February). Rainwater Harvesting: A Comprehensive Review of Literature. Retrieved from https://repository.lib.ncsu.edu/bitstream/handle/1840.4/8170/1_NC-WRRI-425.pdf

Gutierrez-Jimenez, J. (2014). Evaluation of a Point-Of Use Water Purification System (Llaveoz) in a Rural Setting of Chiapas, Mexico. Journal of Microbiology & Experimentation, 1(3). doi:10.15406/jmen.2014.01.00015

Haan, C. T., Barfield, B. J., & Hayes, J. C. (1999). Design Hydrology and Sedimentology for Small Catchments. San Diego u.a.: Academic Press

Handwashing: Clean Hands Save Lives. (2015, November 18). Retrieved from https://www.cdc.gov/handwashing/why-handwashing.html

Hazeltine, B.; Bull, C. (1999). Appropriate Technology: Tools, Choices, and Implications. New York: Academic Press. pp. 3, 270. ISBN 0-12-335190-1

Kniffen, B., Clayton, B., Kingman, D., Jaber, F. (2012). Rainwater Harvesting: System Planning. Fort Stockton, TX. Texas A&M University. Pg 71.

Maldonado, Iván Nelinho Pérez, Lucia Guadalupe Pruneda Álvarez, Fernando Díaz-Barriga, Lilia Elisa Batres Esquivel, Francisco Javier Pérez Vázquez and Rebeca Isabel Martínez Salinas (2011). Indoor Air Pollution in Mexico, The Impact of Air Pollution on Health, Economy, Environment and Agricultural Sources, Dr. Mohamed Khallaf (Ed.), InTech, DOI: 10.5772/19864. Also retrieved from https://www.intechopen.com/books/the-impact-of-air-pollution-on-health-economy-environment-and-agricultural-sources/indoor-air-pollution-in-mexico

Martinson, B., & Thomas, T. (2005). Quantifying the first flush phenomenon. In 12th International Rainwater Catchment Systems Conference

Mendez, C. B., Afshar, B. R., Kinney, K., Barrett, M. E., & Kirisits, M. (2010, January). Effect of Roof Material on Water Quality for Rainwater Harvesting Systems. Retrieved from https://greywateraction.org/wp-content/uploads/2014/11/Effect-of-Roof-Material-on-Water-Quality-for-Rainwater-Harvesting-Systems.pdf

National Primary Drinking Water Regulations. (2017, March 21). Retrieved from https://www.epa.gov/ground-water-and-drinking-water/national-primary-drinking-water-regulations

Pearce, Joshua M., Lonny Grafman, Thomas Colledge, Ryan Legg, (2008). Leveraging Information Technology, Social Entrepreneurship, and Global Collaboration for Just Sustainable Development. 12th Annual NCIIA, 201-210

Porter, D.O., Persyn, R.A., Silvy, V.A. (2004). Rainwater Harvesting. Fort Stockton, TX. Texas A&M University.

Pressure Units - Online Converter. (n.d.). Retrieved from http://www.engineeringtoolbox.com/pressure-units-converter-d_569.html

Rainfall on Roofs and Gutter Slopes. (n.d.). Retrieved from http://www.engineeringtoolbox.com/sloopes-roof-drainage-d_1107.html

Texas Commission on Environmental Quality (2007). Harvesting, Storing, and Treating Rainwater for Domestic Indoor Use. Retrieved from http://rainwaterharvesting.tamu.edu/files/2011/05/gi-366_2021994.pdf

Waterfall, P. (2006). Harvesting Water for Landscape Use. Retrieved from https://extension.arizona.edu/sites/extension.arizona.edu/files/pubs/az1344.pdf

Yujie, Q., De Gouvello, B., & Bruno, T. (2016, June). Qualitative characterization of the first-flush phenomenon in roof-harvested rainwater systems. In LID (low impact development conference) 2016.

Index

Author Bio

Lonny Grafman is an Instructor at Humboldt State University; the founder of the Practivistas summer abroad, full immersion, resilient community technology program; the project manager of the epi-apocalyptic city art project Swale; the Chief Product Officer of Nexi; and the President of the Appropedia Foundation, sharing knowledge to build rich, sustainable lives.

Lonny has developed courses at universities in four countries and facilitated engagements around the world. He has worked, and led teams, on hundreds of domestic and international projects across a broad spectrum of sustainable design and entrepreneurship – from solar energy to improved cookstoves, from micro-hydro power to rainwater catchment, from earthen construction to plastic bottle schoolrooms. Throughout all these technology implementations, he has found the most vital component to be community.

TWITTER	LINKEDIN	FACEBOOK	INSTAGRAM
@lonnygrafman	LonnyGrafman	Lonnyg	@lonnygrafman

Made in the USA
Columbia, SC
23 August 2018